T0073163

Lecture Notes on Engineering Human Thermal Comfort

Lecture Notes on Engineering Human Thermal Comfort

David S-K Ting

University of Windsor, Canada

World Scientific

NEW JERSEY · LONDON · SINGAPORE · BEIJING · SHANGHAI · HONG KONG · TAIPEI · CHENNAI · TOKYO

Published by

World Scientific Publishing Co. Pte. Ltd.

5 Toh Tuck Link, Singapore 596224

USA office: 27 Warren Street, Suite 401-402, Hackensack, NJ 07601

UK office: 57 Shelton Street, Covent Garden, London WC2H 9HE

Library of Congress Control Number: 2020932214

British Library Cataloguing-in-Publication Data
A catalogue record for this book is available from the British Library.

LECTURE NOTES ON ENGINEERING HUMAN THERMAL COMFORT

ISBN 978-981-120-174-5

For any available supplementary material, please visit
https://www.worldscientific.com/worldscibooks/10.1142/11318#t=suppl

Desk Editor: Amanda Yun

Dedication

Luxury and complacency are probably neither good for our environment nor our soul. Thermal comfort, on the other hand, is essential for human survival and proper functioning. This book is dedicated to those who strive to sustain human thermal comfort for the next generations, creating an environment to foster positive progress in a sustainable manner.

"Concern for man and his fate must always form the chief interest of all technical endeavors. Never forget this in the midst of your diagrams and equations."

– Albert Einstein

Contents

Preface

This book is primarily meant for senior undergraduate engineering students interested in engineering human thermal comfort. It invokes some undergraduate knowledge of thermodynamics, heat transfer, and fluid mechanics. This background is furnished as needed to enable every keen being to appreciate thermal comfort engineering. After all, every human being leans on thermal comfort to thrive.

David S-K. Ting
July 31, 2019

Acknowledgments

Completing such a book is indeed a marathon undertaking. With less than the required stamina, the author leans on many good hands to carry him through, at times in a stretcher, and cross the finish line. Notably, grace and strength lavished daily from above have kept him cool when the striving was hot, and warm when it was frigid. The following individuals have been instrumental in realizing this dream of the author, engineering thermal comfort for everyone.

Prof. Dr. Thomas W. McDonald, who brought the back-then completely inexperienced freshman faculty member with next-to-zero HVAC&R background into teaching Air Conditioning. Thankfully, this senior undergraduate course is but the beautiful application of engineering thermofluids for human thermal comfort. This beauty has since been one of the most favored courses the author savors.

How could a mind be opened into engineering thermal comfort if it cannot feel and visualize the hot, the cold, and the cool? The Turbulence and Energy (T&E) Laboratory (http://www.turbulenceandenergylab.org/) forges not only engineering enthusiasts into experts, but also engenders many artistic engineering experts. These skillful T&E experts are credited in the corresponding figure captions. Their workmanship is explicitly displayed throughout the book, where the particular engineering artist is recognized in the figure caption.

The many T&E forged eagle eyes who picked up big, small, and ugly errors scattered throughout the manuscript. A heartfelt Thank You goes to my T&E colleague, Dr. Jacqueline Stagner, who suffered through the working drafts, from the first to the last word.

The many uncanny humors scattered in the book are more-or-less the byproduct of a genetic tract nurtured by the arcane environment. Thanks to mom, dad, sisters, brother, and the enchanting rainforest of Borneo, where thermal comfort for the orangutans is provided by the rich natural canopy and opulent afternoon thunder showers. The Allinterest Research Institute has bridged the suspense between the mesmerizing orangutan habitat to the thought-provoking Turbulence and Energy Laboratory.

Naomi, Yoniana, Tachelle and Zarek Ting for furnishing a 'thermally-comfortable' environment with their unfailing love. They also provided occasional aberrations from the serene state, breaking up any monotonous writing. Thanks for showering the author spasmodically with sarcasms, motivations in disguise, all of which contributed to the realization of an otherwise elusive daydream.

CHAPTER 1

What is Thermal Comfort?

"You can be like a thermometer, just reflecting the world around you. Or you can be a thermostat, one of those people who sets the temperature."

–Cory Booker

Nomenclature

ACH	Air change per hour
ARI	Air conditioning and Refrigeration Institute
ASHRAE	American Society of Heating, Refrigeration and Air Conditioning Engineers
ASHVE	American Society of Heating and Ventilating Engineers
ASRE	American Society of Refrigeration Engineers

(Continued)

HVAC	heating, ventilation, and air conditioning
HVAC&R	heating, ventilation, air conditioning, and refrigeration
IAQ	indoor air quality
PMV	predicted mean vote
PPD	predicted percentage of dissatisfied (population)

1.1 INTRODUCTION

Are you comfortable? When your state of mind and physical body are at ease, i.e., with the body heat generation in equilibrium with the heat loss to the surroundings, you would likely feel comfortable; see Fig. 1.1. HVAC (heating, ventilation, and air conditioning) is the technology for providing occupant thermal comfort, sheltering one from the unpleasant outdoor environment. Or, more correctly, HVAC is meant to ensure denizens thermal comfort in the midst of the highly-fluctuating weather. As it focuses on supplying and maintaining an indoor environment that is thermally comfortable to the occupants, indoor temperature and humidity are two deciding parameters that need to be regulated. Almost universally, the comfortable indoor temperature falls between 20°C and 25°C, with a corresponding relative humidity in the range of 40–60%. Depending on many other determinants, these ranges of conditions furnish the befitting heat sink for removing the right quantity of heat per unit time from our body, maintaining it comfortably at 37°C. Other than temperature and humidity, some minimum air movement is needed for appropriate heat convection and for supplying adequate fresh air to the occupant; see Fig. 1.2. Furthermore, lighting and backdrop, including background melody and scenery, etc., are also necessary to sooth the soul.

Historically, the quest for more comfortable living near the North Pole can be clearly seen from the ingenious creation of Igloos from snow by the indigenous inhabitants; see Fig. 1.3. The very representation of coldness, snow, is surprisingly an excellent insulator. Therefore, there is a certain amount of truth in the Chinese proverb, "fighting poison with poison." The parallel saying in North America is "fighting fire with fire." In heating, ventilation, and air conditioning (HVAC) context, the indigenous people fight cold with cold, or, more correctly, snow. Igloos not only resist the outdoor coldness from penetrating indoors via conduction but also retain the occupant-generated heat indoors. The highly-reflective snow also keeps radiation heat transfer in check. Moreover, the design of the igloo is such that the

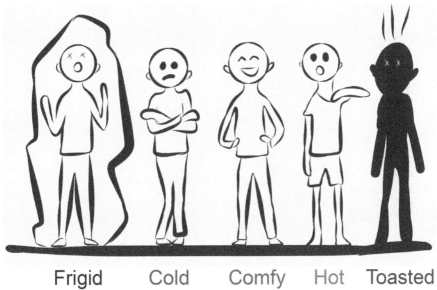

Frigid   Cold   Comfy   Hot   Toasted

Figure 1.1. Thermal comfort illustrated (created by S. Akhand).

Figure 1.2. Minimum air movement for thermal comfort (created by T.A. Tirtha).

Figure 1.3. Fighting cold with snow (created by S. Akhand).

portal is positioned away from the prevailing wind and, hence, minimizes convection heat loss. In short, an igloo keeps out the wind, the snow, and the cold. It keeps you warm.

Other than the igloo that is built from naturally-available snow near the North Pole, inhabitants farther away from the poles have traditionally resorted to fireplaces to beat the long, cold winter. Closer to the equator, on the other hand, trying to stay cool has been a more important striving. It is interesting to note that the first air conditioning units operated by passing warm air over an array of ice blocks [ASHRAE, 1999].

The energy (oil) crisis of the 1970s led to the emerging of well-sealed and highly-insulated buildings, especially in North America. Later in the 1980s, the indoor air quality (IAQ) became a concern. Consequently, minimum air exchange between the indoor stale air and outdoor fresh air came into place. For larger buildings, this is ensured via central air distribution systems, with or without cooling or heating capability. It is interesting to note that we are yet to figure out exactly what is needed for a healthy indoor environment, namely, the value of the minimum required air exchange is still being debated and altered every now and then. Nonetheless, the rule of thumb is approximately 0.5 ACH[1] (air change hour).

Undoubtedly, ASHRAE (the American Society of Heating, Refrigerating and Air Conditioning Engineers) is all over the map on HVAC and/or thermal comfort. Recently, ASHRAE, having been a global, beyond just American, professional association for many years since its inception, is trying to dissociate itself from the original terms making up the acronym. Along this effort to make the first letter of the acronym, A, not signifying American, ASHRAE is thus declared to stand for nothing. Even though ASHRAE no longer stands for the American Society of Heating, Refrigerating and Air Conditioning, its mandate, nevertheless, remains at advancing heating, ventilation, air conditioning, and refrigeration, and, if the author may add, for human thermal comfort. Concerning the establishment of ASHRAE, the following milestones are worth highlighting.

1894 Hugh J. Barron founded the American Society of Heating and Ventilating Engineers (ASHVE).
1904 Refrigeration engineers formed the American Society of Refrigeration Engineers (ASRE).
1954 ASHVE and ASRE merged to become ASHRAE.

Among others, also founded was ARI (the Air conditioning and Refrigeration Institute). ARI is a national trade association representing manufacturers of over 90% of United-States-produced central air conditioning and commercial refrigeration equipment.

Back to the question of thermal comfort, conventionally, predicted mean vote (PMV) index has been employed to predict the mean response of occupants according to the ASHRAE thermal sensation scale; see Chapter 9 of *ASHRAE 2017 Handbook: Fundamentals* [ASHRAE, 2017]. One PMV scale is depicted in Fig. 1.4. Keep in mind that the neutral or comfortable PMV value of zero falls around the neighborhood of 22°C and 50% relative humidity. The word "predicted" here relates to the large number (more than one thousand) of "guinea pigs," originally all college students, who participated in climate chamber experiments, from which statistical votes were deduced to represent the general population at large. Deviations from

[1]ACH signifies the air change rate, i.e., the rate of replacing the indoor air with outdoor air, in terms of volume of the building of interest per unit hour.

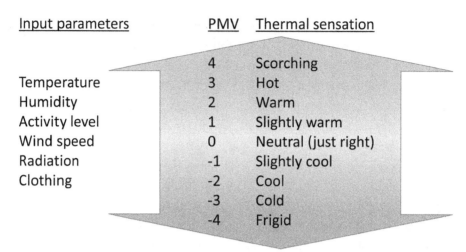

Input parameters	PMV	Thermal sensation
	4	Scorching
Temperature	3	Hot
Humidity	2	Warm
Activity level	1	Slightly warm
Wind speed	0	Neutral (just right)
Radiation	-1	Slightly cool
Clothing	-2	Cool
	-3	Cold
	-4	Frigid

Figure 1.4. Thermal comfort rating and predicted mean vote (PMV) (created by D. Ting). Temperature, humidity, occupant activity level, local air flow, radiation, and clothing are key influential parameters affecting PMV.

either side of the neutral comfortable condition (temperature) are predicted to lead to a progressively larger percentage of the population being dissatisfied. This is illustrated in Fig. 1.5 in terms of predicted percentage of dissatisfied (PPD) versus PMV, where 100% of the population is predicted to be dissatisfied when it is too cold or too hot. As expected, it is impossible to satisfy every person with one comfortable setting. Admittedly, this, in part, is because of some amount of variation in personal physiology; certain people are more "warm-blooded" or "hot," whereas some tend to be "cold-blooded" or "cool." To irritate everybody, on the other hand, is very doable. Pushing the thermostat setting adequate up or down will aggravate the entire population. This can be seen in Fig. 1.5, which indicates that the PPD quickly approaches 100 when it is very cold or very hot.

It should be mentioned that there has been a surge of newer publications of thermal comfort assessments in recent years. Cheung et al. [2019] found that the accuracy of PMV-PPD, when applied to the entire population in the ASHRAE Global Thermal Comfort Database II [Földváry Ličina et al., 2018], is only 34%, as compared to 43% accurate when relating occupant's thermal sensation (satisfaction) to the air temperature alone. This is somewhat perplexing because four environmental parameters, air temperature, mean radiant temperature, relative humidity, and air speed, and two personal parameters, metabolic rate and clothing insulation, are involved in PMV. One key reason why the PMV-PPD approach failed to provide a better prediction is this classical study conducted by Fanger [1970] was based on a

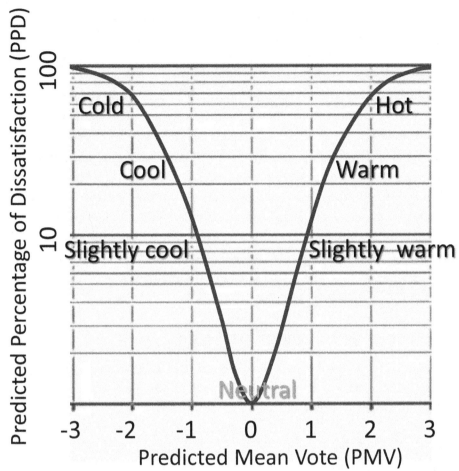

Figure 1.5. Predicted percentage of dissatisfied (PPD) versus predicted mean vote (PMV) (created by D. Ting). Note that there is a small percentage of human species that have peculiar preferences when it comes to thermal comfort. These anomalies, nonetheless, are bounded by the cold and hot limits of ±3 in terms of PMV.

group of participants who were all college students from a temperate climate. The ASHRAE Global Thermal Comfort Database II [Földváry Ličina et al., 2018], on the other hand, consists of a much more diverse and significantly bigger data set that crosses age, culture, climate, etc. For instance, an older person who has spent his entire life in a hot and humid tropical climate will likely feel more comfortable in a warmer and more humid environment surrounded by vegetation than a young-ster from a frigid climate/zone who enjoys playing in the snow. Also worth men-tioning is that the range of participant activities covered in Fanger's experiment is quite limited. Wang and Hu [2018] found that people start to sweat while undergoing moderate activities, and their mean thermal sensation vote tends to increase. In other words, the level of physical, mental, and spiritual activities can noticeably alter the thermal comfort settings.

There is definitely some weighty amount of art and subjectivity, within the science of thermal comfort. To cite but one example is the debate disclosed by Chappells and Shove [2005], which eludes to the notion that thermal comfort is a highly negotiable social–cultural construct. An analogous everyday example is buffet. It is an universal challenge to not overeat at a buffet, where we are tempted with all the food choices. On the other hand, a healthy banquet has the set amount of food, and in general, we are more satisfied after a quantity-controlled banquet than a buffet. In the same token, we can alter the unsustainable trajectories of the need and technologies associated with thermal comfort into sustainable ones. Adaptive thermal comfort [Carlucci et al., 2018] is one probable undertaking along the path to more sustainable HVAC.

1.2 THE ROLE OF AN HVAC&R ENGINEER

What does an HVAC&R engineer do? Being compensated for solving heating and cooling problems may be a short answer. The air conditioning fable given by Pita [1998] is timely. Here is a paraphrase:

On a record-breaking summer day, the air conditioning system in a skyscraper stopped functioning, turning the offices into steam saunas. Without operable windows, computers began to break down, office workers started to leave, and tenants threatened lawsuits for damages. Not knowing what to do, the managerial team became frantic. Just when the world seemed to collapse on them, someone shouted, "There is this HVAC wizard going by the name Rupp just a block away." In desperation, they called Rupp. Within minutes, Rupp showed up and briefly examined the complex 7000-ton HVAC system and muttered, "Hmm." He took out a small wrench and tapped a valve. Immediately the system started running and soon thermal comfort was restored. The building manager thanked Rupp and asked him how much he owed him. "$3705," Rupp responded. "Are you nuts?" the manager shrieked, "$3705 for beating up a valve?" "The tapping costs only $5," Rupp responded, "The $3700 is for knowing which valve to tap."

We can see from the above that knowledge is money! But, what about Dilbert's salary theorem? Let us take a little break and savor Example 1.1.

EXAMPLE 1.1. KNOWLEDGE, POWER, TIME, MONEY, AND WORK

Given: Knowledge, power, time, money, and work.
Find: Deduce the appropriate relationships between them.

Solution:

$$\text{Knowledge} = \text{Power} \tag{1.1.1}$$

$$\text{Time} = \text{Money} \tag{1.1.2}$$

$$\text{Power} = \text{Work/Time} \tag{1.1.3}$$

Substitute Eqs. 1.1.1 and 1.1.2 into Eq. 1.1.3,

$$(\text{Knowledge}) = \text{Work/(Money)} \tag{1.1.4}$$

or

$$\text{Money} = \text{Work/Knowledge} \tag{1.1.5}$$

We see that the more one works, the more money one earns. Somewhat surprisingly, the expression also hints that the gaining of knowledge leads to a decrease in one's earning, i.e., money. Cynical as it is, there is a certain amount of truth that furthering one's knowledge beyond that needed to make a living can indeed bring about a reduction in one's income.

Returning to the question concerning the role of HVAC engineers, in general, the tasks of HVAC engineers are to calculate the demands for heating, cooling, and ventilation; to choose the necessary equipment and controls; and to ensure that the components are correctly integrated into the building. As mentioned earlier, thermal comfort depends on:

Temperature. Both dry-bulb and mean-radiant temperatures are important. The dry-bulb temperature more-or-less dictates the convection heat transfer between our body and the surrounding air. On the other hand, our skin and the outer surface of our clothing exchange heat with the ambient surfaces via radiation.

Humidity. A relative humidity in the ballpark of 50% is a good guideline.

Air Motion. Near-stagnant air is never a comfortable environment. Respiration, body odor diffusion, burping, and tooting are all healthy bodily functions. Some of these may amplify when our body endeavors to adjust to become more comfortable with its environment. Some minimum air movement is needed to convect and diffuse these chemicals to ensure proper functioning of our mind.

Airborne Contaminants. Both gaseous species and particulate need to be kept in check. The many happenings indoor such as the usage of a multitude of beauty sprays can adversely impact the IAQ.

State of Mind. Individuals such as the spiritual gurus and world-class magicians can control their mind to a large extent. The lowering of their heart beat leads to reduced metabolism followed by heat generation, and hence, thermal comfort condition. Some magicians can convince their mind that they are on a warm sunny beach when they are enclosed in ice.

The HVAC design process involves basically iterating the following steps:

Calculation of peak loads.
Specification of equipment and system configuration.
Calculation of annual performance.
Calculation of costs.

Note that for situations such as those of a hospital, museum, or a pharmaceutical firm, special and unique attentions are required.

EXAMPLE 1.2. LIFE CYCLE COST, THE ECONOMICS OF ENERGY EFFICIENCY

Given: Two air-conditioners provide 20 years of quality service. Model Eco costs $3000 and is powered at 200 W. Model Eff is $3200 and uses 160 W. The price of electricity is 12¢/kWh.

Find: Deduce the better choice, based on life cycle analysis.

Solution:

The difference in annual cost is

$$8760\,\text{h} \times (0.20 - 0.16)\,\text{kW} \times \$0.12/\text{kWh} = \$42.05 \text{ annual savings.}$$

Note that there are 8760 h in a year.
Neglecting inflation, interest, etc.,

$$\text{Payback time} = \text{investment/annual savings} = \$200/\$42.05 = 4.8\,\text{years.}$$

The slightly more expensive model Eff is a great deal, considering the 20-year service life.

The profitability is generally considered excellent if the payback time is within one-third of the lifetime. It is good if the payback time is no more than one-half of the lifetime.

1.2.1 Why Do We Bother with this Low-Tech Field of HVAC&R?

The human species is not very adaptive, and the situation may be getting progressively worse, especially with the progressive entitlement mentality along with climate change. We need individual air-conditioning adjustment units even when we travel over short distances in a minivan! It cannot just be 21°C with a little breeze for everybody anymore. On the flip side, depending on the day and the mood of the person, a regular sauna visitor could be the very one who insists on a comfortable temperature of no more than 18°C with 70% or higher relative humidity! The bottom line is that HVAC&R is everywhere!

One can get a sense of the ever-importance of this field from the review on human thermal comfort in the built environment by Rupp et al. [2015]. Keep in mind that globally more than 40% of total building energy is spent in thermal comfort, i.e., totally some 49,000 PJ in 2016 [IEA, 2017]. This ratio and more so, the absolute quantity, will escalate as developing countries advance into widespread indoor thermal comfort engineering. In addition, climate change is expected to add a heavy load to the cooling demand.

1.3 ORGANIZATION OF THE BOOK

It is worth listing some of the main textbooks on air conditioning. Each of these has its strengths and shortcomings. The appropriateness depends on the particular need. The list is naturally biased by the author's familiarity, i.e., he has adopted or at least read parts of the following books. Some of these books have since been updated with newer editions.

Air-Conditioning and Refrigeration Institute, *Refrigeration and Air Condition-ing*, 3rd ed., Prentice Hall, Upper Saddle River, 1998.

ASHRAE, *2017 ASHRAE Handbook: Fundamentals*, 2017.

J.F. Kreider, P.S. Curtiss, A. Rabl, *Heating and Cooling of Buildings: Design for Efficiency*, 2nd ed., McGraw-Hill, New York, 2002.

T.H. Kuehn, J.W. Ramsey, J.L. Threlkeld, *Thermal Environmental Engineering*, 3rd ed., Prentice Hall, Upper Saddle River, 1998.

F.C. McQuiston, J.D. Parker, J.D. Spitler, *Heating, Ventilation, and Air Condi-tioning: Analysis and Design*, 6th ed., Wiley, Hoboken, 2005.

J.W. Mitchell, J.E. Braun, *Principles of Heating, Ventilation, and Air Conditioning in Buildings*, Wiley, Hoboken, 2013.

E.G. Pita, *Air Conditioning Principles and Systems*, 3rd ed., Prentice Hall, Upper Saddle River, 1998.

N.E. Wijeysundera, *Principles of Heating, Ventilation and Air Conditioning with Worked Examples*, World Scientific, Singapore, 2016.

This book takes advantage of ASHRAE handbooks, especially Fundamentals, for the ever-accumulating data needed to properly perform the involved HVAC calculations and analyses. With the authoritative complement, it overcomes a com-mon drawback of many existing air conditioning textbooks, which contain incom-plete data. An important fringe benefit is the unnecessary thickening of the book with incomplete resources. *Lecture Notes on Engineering Human Thermal Com-fort* aims at explaining the quintessence of engineering human thermal comfort in the student-friendly and attention-seizing Çengel style. Straight-forward writ-ing is interjected with illustrative figures, anecdotic banter, and ironical analogies, to facilitate the comprehension of the materials and to provide timeous stimuli in the midst of arduous technical details. As the objective of this book is to furnish the essentials for appreciating HVAC engineering, some details are intentionally excluded to prevent them from muddying the basics.

The thermodynamics underlying HVAC are reviewed in Chapter 2. Chapter 3 expounds on the study of moist air, i.e., psychrometry. With that backdrop, Chapter 4 details various processes with the help of the psychrometric chart. Heat transmission in a building is disclosed in Chapter 5. Subsequently, the largely solar-less heating in the winter is elucidated in Chapter 6. Solar radiation is necessarily covered in Chapter 7, before one can deal with the heavily sun-dependent cooling in the summer, Chapter 8. Money talks, and hence, the energy usage needs to be accurately estimated, and this is illustrated in Chapter 9. Selected heating and cooling systems are conveyed in Chapter 10. Much can be learned from the intelligent design displayed in nature. This is touched-on in Chapter 11, with the intention to stir up creative young minds into engineering more excellent HVAC systems and designs.

PROBLEMS

For this introductory chapter, answering the general questions below requires the scholar to read and backup the answers with facts started in academic articles, especially those published in recent issues of the standard journals. Chen et al. [2018], D'Oca et al. [2018], and Tian et al. [2018] are three sample articles that may be of use.

Problem 1.1

Approximately what percentage of world energy usage is for providing human thermal comfort indoor? Find a region or country where this percentage, with respect to the total regional or national energy usage, is the highest. Explain the reasons behind this. Support your answers with reliable published work.

Problem 1.2

From the literature, roughly how much time does a typical person stay indoors? How does this vary with climate zone?

Problem 1.3

Choose the most appropriate answer concerning igloos.
Igloos are made of snow because snow is a good

  (A) thermal insulator
  (B) thermal conductor

(C) thermal capacitor

(D) solar collector

The emissivity of snow is

(A) high

(B) moderate

(C) low

and thus also

(A) conduction

(B) convection

(C) radiation

heat transfer.

Problem 1.4

Based on recently published work, how does thermal comfort satisfaction vary with occupant density?

Problem 1.5

What does recent research say concerning the productivity of an individual with respect to her/his thermal comfort?

Problem 1.6

Find the literature which studies the uncertainty associated with the assessment of building energy usage. How can one improve the assessment?

Problem 1.7

Many recent studies are saying that occupant behavior has a significant impact on building energy usage, and that this contributes greatly to the discrepancy between estimated and actual building energy usage. Explain.

REFERENCES

Air-Conditioning and Refrigeration Institute, Refrigeration and Air Conditioning, 3rd ed., Prentice Hall, Upper Saddle River, 1998.

American Society of Heating, Refrigeration and Air Conditioning Engineers (ASHRAE), *2017 ASHRAE Handbook: Fundamentals*, 2017.

S. Carlucci, L. Bai, R. De Dear, L. Yang, "Review of adaptive thermal comfort models in built environmental regulatory documents," Building and Environment, 137: 73–89, 2018.

H. Chappells, E. Shove, "Debating the future of comfort: environmental sustainability, energy consumption and the indoor environment," Building Research & Information, 33(1): 32–40, 2005.

Y. Chen, H. Liu, L. Shi, "Operation strategy of public building: implications from trade-off between carbon emission and occupant satisfaction," Journal of Cleaner Production, 205: 629–644, 2018.

T. Cheung, S. Schiavon, T. Parkinson, P. Li, G. Brager, "Analysis of the accuracy on PMV-PPD model using the ASHRAE global thermal comfort database II," Building and Environment, 153: 205–217, 2019.

S. D'Oca, T. Hong, J. Langevin, "The human dimensions of energy use in buildings: a review," Renewable and Sustainable Energy Reviews, 81: 731–742, 2018.

P.O. Fanger, *Thermal Comfort: Analysis and Application in Environmental Engineering*, Danish Technical Press, Copenhagen, 1970.

V. Földváry Ličina et al "Development of the ASHRAE global thermal comfort database II," Building and Environment, 142: 502–512, 2018.

International Energy Agency (IEA), *Market Report Series: Energy Efficiency 2017*, 2017.

J.F. Kreider, P.S. Curtiss, A. Rabl, *Heating and Cooling of Buildings: Design for Efficiency*, 2nd ed., McGraw-Hill, New York, 2002.

T.H. Kuehn, J.W. Ramsey, J.L. Threlkeld, *Thermal Environmental Engineering*, 3rd ed., Prentice Hall, Upper Saddle River, 1998.

B. Nagengast, "Comfort from a block of ice: a history of comfort cooling using ice," ASHRAE Journal, 41(2): 49–57, 1999.

F.C. McQuiston, J.D. Parker, J.D. Spitler, *Heating, Ventilation, and Air Conditioning: Analysis and Design*, 6th ed., Wiley, Hoboken, 2005.

J.W. Mitchell, J.E. Braun, *Principles of Heating, Ventilation, and Air Conditioning in Buildings*, Wiley, Hoboken, 2013.

E.G. Pita, *Air Conditioning Principles and Systems*, 3rd ed., Prentice Hall, Upper Saddle River, 1998.

R.F. Rupp, N.G. Vásquez, R. Lamberts, "A review of human thermal comfort in the built environment," Energy and Buildings, 105: 178–205, 2015.

W. Tian, Y. Heo, P. de Wilde, Z. Li, D. Yan, C.S. Park, X. Feng, G. Augenbroe, "A review of uncertainty analysis in building energy assessment," Renewable and Sustainable Energy Reviews, 93: 285–301, 2018.

H. Wang, S. Hu, "Analysis on body heat losses and its effect on thermal sensation of people under moderate activities," Building and Environment, 142: 180–187, 2018.

N.E. Wijeysundera, *Principles of Heating, Ventilation and Air Conditioning with Worked Examples*, World Scientific, Singapore, 2016.

CHAPTER 2

HVAC Thermodynamics

"Classical thermodynamics... is the only physical theory of universal content which I am convinced... will never be overthrown."

–Albert Einstein.

CHAPTER OBJECTIVES

- Recap relevant thermodynamics pertinent to heating, ventilation, and air conditioning (HVAC).
- Invoke conservation of energy and mass to an open HVAC system.
- Differentiate sensible heat from latent heat.
- Recognize pressure as flow energy.
- Apply the first law of thermodynamics in terms of enthalpy to a steady-flow open system.

Nomenclature

c_p	heat capacity at constant pressure; $c_{p_{air}}$ is the heat capacity of air
COP	coefficient of performance; COP_{HP} is the coefficient of performance of a heat pump
CV	control volume
d	change
E	energy
e	energy per unit mass; e_{mech} is mechanical energy per unit mass
g	gravity or gravitational acceleration

(Continued)

h specific enthalpy; h_f is the specific enthalpy of the liquid phase, h_{fg} is the difference in specific enthalpy between the gaseous phase and the liquid phase, h_g is the specific enthalpy of the gaseous phase, $h_{steam,i}$ is the specific enthalpy of the incoming steam, $h_{steam,o}$ is the specific enthalpy of the outgoing steam

H_2O water, one oxygen atom bonded with two hydrogen atoms

HEX heat exchanger

HV heating value

HVAC heating, ventilation, and air conditioning

m mass; m' is mass flow rate, m_{air}' is mass flow rate of air, m_{steam}' is the mass flow rate of steam

P pressure

Q heat; Q_{air}' is the heat transfer rate of air, Q_H is the heat transfer to a higher temperature space, Q_{in}' is the heat transfer rate into the system, Q_{rqd}' is the required heat input rate, Q_{steam}' is the heat transfer rate of steam

SI International system of units

T temperature; $T_{air,i}$ is incoming air temperature, $T_{air,o}$ is outgoing air temperature, T_i is indoor temperature, T_o is outdoor temperature, $T_{steam,i}$ is the incoming steam temperature, T_{sat} is the saturation temperature

t time

u specific internal energy

V velocity

v specific volume; v_f is the specific volume of the liquid phase, v_{fg} is the difference in specific volume between the gaseous phase and the liquid phase, v_g is the specific volume of the gaseous phase

W work; W_{in} is the work input, W_{in}' is the rate of work input

z vertically upward direction

Greek and Other Symbols

Δ change

ε_{HEX} effectiveness of a heat exchanger

η efficiency; η_{comb} is the combustion efficiency

ρ density; ρ_{air} is the density of air

∀ volume; $\forall'$ volume flow rate

2.1 INTRODUCTION

This chapter aims at recapping the essential elements of thermodynamics underlying HVAC. For a more detailed recapitulation, one can refer to Çengel and Boles [2015] and Borgnakke and Sonntag [2017]. These fundamentals are also reviewed in HVAC books such as Kreider et al. [2002], Kuehn et al. [1998], Mitchell and Braun [2013], and McQuiston et al. [2005].

What is **thermodynamics**? Within the context of HVAC, thermodynamics deals with energy, its various forms, transformation, and conservation. Of particular relevance is energy in thermal form, commonly referred to as heat, which is veritably thermal energy in transit. An obvious case is the thermal energy balance of a building, as depicted in Fig. 2.1. During the winter time, the heat loss from a building to the outdoor minus the total heat gained from the sun and internal sources must be supplied by the heating system, to ensure livability. As such, the main form of energy of concern here is the thermal energy, along with its conservation.

The most common thermodynamic system involved in this book is the building itself. A building is an **open system**[1] where both energy and mass can cross its

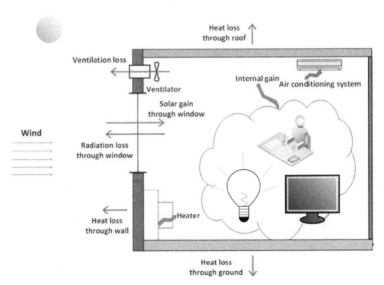

Figure 2.1. Thermal energy balance of a building (created by X. Wang). Under steady-state conditions, the sum of all the heat loss is equal to total heat gain.

---

[1]A system is a domain of interest. In contrast to a **closed system**, an **open system** also allows mass, in addition to energy, to cross the boundary of the system.

boundary; see Fig. 2.2. The thermal energy crosses a building most commonly via radiation through fenestration, convection–conduction–convection across walls and roofs, and also contained within and transported via the moving air mass through various openings. Concerning the air mass of a building, **conservation of mass** is realized as exfiltration (conditioned air leaving the building) is balanced by infiltration (unconditioned air entering the building).

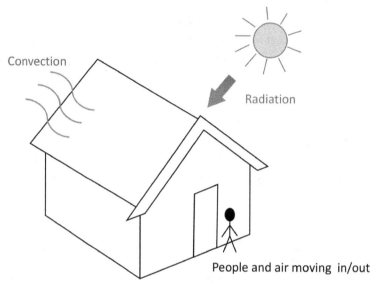

Figure 2.2.   The building as an open thermodynamic system (created by Y. Yang). It is an open system because both energy and mass can enter or exit the boundary of the building.

EXAMPLE 2.1. AIR HEATING VIA A NON-MIXING HEAT EXCHANGER (HEX) WITH FLOWING STEAM

Given: Atmospheric air at 20°C flowing at 1 m³/s in a duct passes through a heat exchanger (HEX) and exits at 50°C as shown in Fig. 2.3. Saturated steam at atmospheric pressure enters the coil of the HEX, and it exits with a 20% quality.

Find: (a) The heat transfer rate from the steam to the air, (b) the mass flow rate of the steam, (c) the HEX effectiveness.

Solution:

The amount of thermal energy gained by the air passing through the HEX:

$$Q_{air}' = m_{air}' c_{P,air} \left(T_{air,o} - T_{air,i}\right) \tag{2.1.1}$$

The mass flow rate of air is equal to the density times the volume flow rate, $m_{air}' = \rho_{air} \dot{V}' = 1.2 (1) = 1.2$ kg/s, the heat capacity of air, $c_{P,air} = 1$ kJ/kg-K, and the outgoing and incoming air temperatures are, respectively, $T_{air,o} = 50°C$, $T_{air,i} = 20°C$. Therefore, the heat transfer rate from the steam to the air, $Q_{air}' = 36$ kW.

On the other side, i.e., the inside of the HEX, the rate of thermal energy loss from the saturated incoming steam:

$$Q_{steam}' = m_{steam}' \left(h_{steam,i} - h_{steam,o}\right) \tag{2.1.2}$$

The enthalpy of the saturated steam, $h_{steam,i} = 2676$ kJ/kg. At 20% quality, the enthalpy of the outgoing steam, $h_{steam,o} = 419.1 + 0.2 (2257) = 870.5$ kJ/kg.

According to the first law of thermodynamics, the energy gained by the air is equal to that lost by the steam, i.e., $Q_{air}' = Q_{steam}' = 36$ kW. Substituting this into Eq. 2.1.2, we get

$$36 = m_{steam}' (2676 - 870.5)$$

or the mass flow rate of steam, $m_{steam}' = 0.02$ kg/s.

The HEX effectiveness:

$$\varepsilon_{HEX} = \left(T_{air,o} - T_{air,i}\right) / \left(T_{steam,i} - T_{air,i}\right) \tag{2.1.3}$$

At atmospheric pressure, saturated steam pertains to a temperature, $T_{steam,i} = 100°C$. Therefore, $\varepsilon_{HEX} = 0.375$.

2.2 PHASE CHANGE, SENSIBLE AND LATENT HEAT

A few points covered in undergraduate thermodynamics have been illustrated in Example 2.1. First, water (H_2O) in vapor phase is partially condensed to liquid phase. The 20% quality steam consists of 20% vapor and 80% liquid. Figure 2.4 depicts how, at atmospheric pressure, water sits comfortably in its liquid phase at 20°C. Water at this thermodynamic state, State 1, is in sub-cooled or compressed

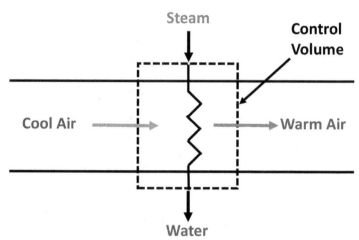

Figure 2.3.   Heating of air via a non-contact heat exchanger with flowing steam (created by S. Akhand).

liquid phase, and it is not about to vaporize. When heating the water in the container with a freely moving piston (constant-pressure heat addition) to approach 100°C, State 2, the liquid water is about to vaporize. Any additional heating will cause the liquid to vaporize. Within the liquid–vapor state, State 3, the two phases of H_2O coexist in equilibrium. The outgoing steam in Example 2.1 falls into this harmonious co-existent state. When the last drop of liquid water vaporizes, the steam inside the constant-pressure container reaches the saturated vapor state, State 4. Any removal of thermal energy will lead to the onset of condensation. The vapor H_2O at State 4 becomes superheated vapor, State 5, with further heating.

Two types of heat are worth noting in Fig. 2.4. Heating the liquid water from State 1 to 2, and also from State 4 to 5, comprises a type of heating we call **sensible heating**. Sensible heating involves energizing the molecules, and thus, the temperature of the medium (H_2O, air) caused by heightened molecule–molecule collisions. On the other hand, **latent heating**, State 3, deals with phase change. It takes significantly more energy to break the H_2O and H_2O attraction[2] as compared to simply causing the molecules to "dance" around more vigorously in sensible

---

[2]In terms of increasing energy intensity or density, the atomic bonds between the oxygen atom and its two hydrogen atoms are substantially harder to break compared to the attractive force between two water molecules. This chemical energy, then again, is exceedingly surpassed by the nuclear energy that involves the "atomic" strong bonds within the nucleus of the atom. If "till death do we part" is the marriage vow, then "nucleus marriages" are the ones that are unbreakable except death.

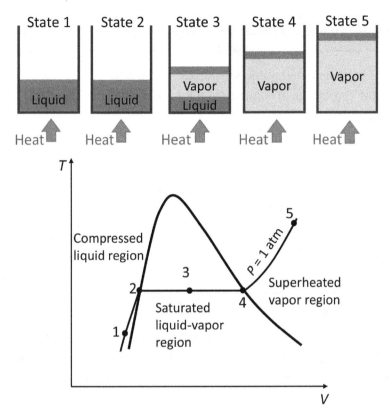

Figure 2.4. Water undergoing phase change at atmospheric pressure (created by Y. Yang, edited by D. Ting).

heating. Therefore, latent heating and cooling is notably more energy intensive, or, in terms of transferring or storing energy, a phase change can provide decidedly denser transport and/or storage capacity. In HVAC, the need of an appropriate level of humidity for thermal comfort, or simply human health, bespeaks the importance of condensation of water vapor and evaporation of liquid water.

In heating and humidification processes using steam, as illustrated in Example 2.1, properties associated with the liquid–vapor equilibrium phase are called upon. Common practice is to use subscript "f" to denote saturated liquid and subscript "g" for saturated vapor. For example, the specific volume (volume per unit mass or mole) of a liquid–vapor mixture,

$$v_{fg} = v_g - v_f \tag{2.1}$$

Similarly, the corresponding (specific) enthalpy,

$$h_{fg} = h_g - h_f \tag{2.2}$$

EXAMPLE 2.2. VOLUME AND ENERGY CHANGE DURING WATER EVAPORATION

Given: 1 kg of saturated water is vaporized at 100 kPa.

Find: (a) The change in volume, (b) the amount of energy required.

Solution:

The volume change per unit mass of water:

$$v_{fg} = v_g - v_f$$

From the thermodynamic table for water, we see that

		Specific volume [m³/kg]		Specific enthalpy [kJ/kg]		
P [kPa]	T_{sat} [°C]	v_f	v_g	h_f	h_{fg}	h_g
100	99.61	0.001043	1.6941	417.51	2257.5	2675.0
100.325	99.97	0.001043	1.6784	419.06	2256.5	2675.6

Therefore, $v_{fg} = v_g - v_f = 1.6941 - 0.001043 = 1.6931$

The total volume change, $\Delta V = m\, v_{fg} = 1\text{ kg} \times 1.6931\text{ m}^3/\text{kg} = 0.3386\text{ m}^3$

The amount of energy required to vaporize the saturated water,

$$\Delta E = m\, h_{fg} = 1\text{ kg} \times 2257.5\text{ kJ/kg} = 2257.5\text{ kJ}$$

2.3 ENERGY CONSERVATION OF AN OPEN SYSTEM

Figure 2.5 depicts a more comprehensive picture regarding energy conservation for an open system. The dashed line defines the **control volume** of the open system of interest. Note that control volume is also used as a synonym for open system. The conservation of energy for the control volume can be expressed as

$$Q_{in}' + W_{in}' + \sum_{in}(m'e) - \sum_{out}(m'e) = d\,(me)_{CV}\,/dt \qquad (2.3)$$

To be precise, the rate of heat and work entering the control volume plus the energy associated with the entering mass minus that contained in the outgoing mass is equal to the rate of energy change of the control volume. For a flowing stream, the energy per unit mass or mass flow rate, the specific energy,

$$e = u + \frac{1}{2}V^2 + gz \qquad (2.4)$$

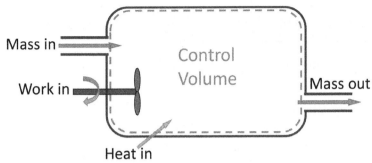

Figure 2.5.   Energy conservation of an open system (created by D. Ting). Work in plus heat in plus energy associated with the mass entering the system is equal to the energy of the exiting mass.

The first term of the right-hand side is the specific **internal energy**. Internal energy is the molecular energy of the system. It is associated with the molecular structure and activity and is comprised of the microscopic energies in terms of translational, vibrational, and rotational kinetic energy of the molecules, as well as the potential energy from intermolecular forces. The second term is the kinetic energy per unit mass of moving fluid, and the last term is the (gravitational) potential energy of the fluid, where V is velocity, g is gravity, and z is elevation.

Pressure is Flow Energy

In SI units, pressure P is in units of

$$\text{Pa (Pascal)} = \text{N/m}^2 = \text{N m/m}^2 = \text{J/m}^3 \tag{2.5}$$

that is, energy (in Joules) per unit volume (cubic meter) of the flowing fluid. Multiplying pressure and the specific volume gives

$$P\forall/m = P/\rho \tag{2.6}$$

It is clear that the SI units for this are J/kg, namely, the product of pressure and specific volume of the moving fluid is energy per unit mass of the fluid. As such, we can recast the specific mechanical energy equation, Eq. 2.4, as

$$e_{\text{mech}} = P/\rho + \frac{1}{2}V^2 + gz \tag{2.7}$$

The first term on the right-hand side is the flow energy, the second, the kinetic energy, and the third, the potential energy, all of which are expressed in terms of per unit mass of the moving fluid.

EXAMPLE 2.3. PRESSURE DROP AND PUMPING REQUIREMENT

Given: The air conditioning system of a space requires 1 kg/s of conditioned air to be delivered.

Find: What can you say about the pumping (fan power) requirement if the air duct is (a) clean and straight, (b) clean with many bends, (c) dirty with many bends?

Solution:

To deliver the same amount of air per unit time, the clean and straight duct requires the least pumping (fan) power.

With many bends, the pressure drop in the air distribution line increases. Therefore, more pumping power is required to deliver the same air flow rate, while overcoming the losses due to the bends.

Dirt adds resistance to the air flow, and hence, the dirty duct with many bands requires the largest pumping power.

When dealing with an open system commonly encountered in HVAC, it is more convenient to use enthalpy instead of internal energy. Specific enthalpy,

$$h = u + Pv \tag{2.8}$$

Here v is the specific volume, $∀/m$. Therefore, we can recast the energy conversation described by Eq. 2.3 as

$$Q_{in}' + W_{in}' + \sum_{in}\left[m'\left(h + \tfrac{1}{2}V^2 + gz\right)\right] - \sum_{out}\left[m'\left(h + \tfrac{1}{2}V^2 + gz\right)\right]$$
$$= d\left[m\left(h + \tfrac{1}{2}V^2 + gz\right)\right]_{CV}/dt \tag{2.9}$$

For the most commonly encountered steady-flow condition, this is reduced into

$$Q_{in}' + W_{in}' + \sum_{in}\left[m'\left(h + \tfrac{1}{2}V^2 + gz\right)\right] - \sum_{out}\left[m'\left(h + \tfrac{1}{2}V^2 + gz\right)\right] = 0 \tag{2.10}$$

EXAMPLE 2.4. STEADY-STATE HEAT BALANCE OF A BUILDING

Given: During a typical day in June, a building in Rotorua, New Zealand, loses 55 kW of thermal energy via the building envelope to the outdoor which is at 8°C. It gains solar energy through the large windows at a rate of 17 kW. Infiltration and exfiltration amount to 3 kg/s of air exchange.

Find: The rate of heating required to maintain the indoor at 20°C.

Solution:

For the assumed steady-state condition, the first law of thermodynamics expressed as

$$Q_{in}' + W_{in}' + \sum_{in}\left[m'\left(h + \tfrac{1}{2}v^2 + gz\right)\right] - \sum_{out}\left[m'\left(h + \tfrac{1}{2}v^2 + gz\right)\right] = 0$$

may be modified into

$$Q_{rqd}' + Q_{in}' - Q_{out}' + \sum_{in}[m'c_pT_o] - \sum_{out}[m'c_pT_i] = 0$$

Here, the first term is the required heat input rate, the fourth term is the rate of thermal energy leaving the building with the outgoing air, exfiltration, and the fifth term is the rate of thermal energy entering the building because of infiltration. Substituting for the given values, we have

$$Q_{rqd}' + 17,000 - 55,000 + 3\,(1000)\,(8 - 20) = 0$$

This gives

$$Q_{rqd}' = 74\,kW$$

2.4 PERFORMANCE INDICATORS

One of the most-utilized performance indicators is grade or grade point average. For example, a student who scores 85% in Engineering Human Thermal Comfort is deemed to be performing well. On the other hand, one that barely makes the 60% mark is underperforming. Analogous to this performance indicator is efficiency, where 100% is the maximum possible efficiency. For keeping a house warm in the winter using a furnace, the efficiency of concern can be defined as the combustion efficiency:

$$\eta_{comb} = Q/HV \tag{2.11}$$

where Q is the desired heat output and HV is the heating value of the fuel, which may be viewed as the thermal energy content of the fuel. As some (heat) losses[3] are inevitable, $\eta_{comb} = 100\%$ is not realized in practice.

Today, the state-of-the-art, high-efficiency furnaces are designed to capture even the heat associated with the condensing exhaust gas. These high-end furnaces can reach an impressive efficiency up to 98%. This is substantially more efficient than conventional furnaces with efficiencies in the range of 70–80%. In point of fact, we do not expect even the finest furnace to perform better than a 98% efficiency. Alternatively, the heating could be accomplished via a **heat pump**. A heat pump is a device which transfers thermal energy from a lower-temperature reservoir, such as the ground or the ambient, to the higher-temperature indoor. This is realized via an inventive device, a heat pump, which operates on a finite amount of work input; see Fig. 2.6. With the heat output, Q_H, as the desired output, and the work input, W_{in}, as the required input, the performance is simply Q_H/W_{in}. This performance indicator is called the **coefficient of performance**, expressly,

$$COP_{HP} = Q_H/W_{in} \tag{2.12}$$

It is noted that COP is equivalent to η, as both signify performance. Nevertheless, η maxes out at unity (100%), while COP can far exceed a value of one. By its very nature, the example above illustrates that the heat pump's efficiency can far exceed 100%, making it the preferred heating system over a furnace when the conditions are right.[4]

2.5 IDEAL GAS

The air that is conditioned for thermal comfort in HVAC is at a low pressure, with respect to its critical pressure, and a high temperature, relative to its critical

---

[3]Also omnipresent is some amount of incomplete combustion. The wise anonymous thermodynamicist summarizes the laws of thermodynamics well: (1) You cannot win, you can only break even. (2) You can only break even at absolute zero. (3) You cannot reach absolute zero. These laws primarily concern the second law of thermodynamics, or entropy. Specifically, (1) the best that we can do is to not generate any entropy, and this can only be achieved at the unachievable absolute zero, i.e., when we are beyond dead. In the present context, it limits the efficiency from reaching 100%.

[4]It is clear that transferring thermal energy from a low-temperature reservoir to a higher-temperature space becomes significantly more difficult as the temperature difference increases. Therefore, heat pumps only make (financial) sense for mild winter regions.

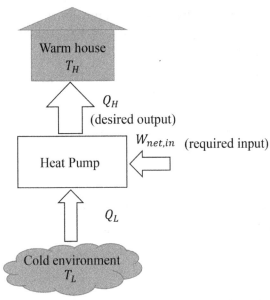

Figure 2.6. Keeping the indoor warm with a heat pump (created by Y. Yang). The amount of thermal energy entering the house is equal to that removed from the cold environment plus the work input.

temperature. The critical point for air is around 37 atm and −140°C. For this reason, it can be assumed to be an ideal gas, viz

$$Pv = RT \qquad (2.13)$$

in literally all HVAC applications. Equation 2.13 is called the ideal-gas equation of state, where P and T are in absolute units. The gas constant,

$$R = R_u/M \qquad (2.12)$$

where the universal gas constant, R_u = 8.314 kJ/kmol·K, and M is the molecular weight of the ideal gas of concern (for air, M = 28.97 kg/kmol).

PROBLEMS

Problem 2.1

A residential furnace has a capacity of 60,000 Btu/h. Find:

(a) The number of 80 W incandescent light bulbs to provide the same amount of heating as the furnace.

(b) The number of florescent light bulbs that provide the same amount of illumination as the number of 80 W incandescent light bulbs in (a).

(c) The number of LED light bulbs that provide the same amount of illumination as the number of 80 W incandescent light bulbs in (a).

Problem 2.2

A 75% efficiency, 2 kW pump is used to pump water over a piping network that has a 100 kPa pressure drop. What is the water flow rate?

Problem 2.3

A school consumes 70 GJ of heat a year. The owner thinks about collecting solar thermal energy in the warm season and storing it in water at 70°C, but usable down to 30°C. What is the required storage tank size?

Problem 2.4

A house runs a 1-ton air conditioner 750 hours through the summer. Find:

(a) The amount of thermal energy removed.
(b) The required electricity, if the COP of the air conditioner is 1.8.
(c) The amount of ice that needs to be gathered in the winter time and stored for this summer cooling.

Problem 2.5

Heat is removed from a house via a 7-kW air conditioner at a rate of 21 kW. What is the COP of the air conditioner? What is the amount of heat rejected to the outdoors?

REFERENCES

C. Borgnakke, R.E. Sonntag, *Fundamentals of Thermodynamics*, 9th ed., Wiley, Hoboken, 2017.

Y.A. Çengel, M.A. Boles, *Thermodynamics: An Engineering Approach*, 8th ed., McGraw-Hill, New York, 2015.

J.F. Kreider, P.S. Curtiss, A. Rabl, *Heating and Cooling of Buildings: Design for Efficiency*, 2nd ed., McGraw-Hill, New York, 2002.

T.H. Kuehn, J.W. Ramsey, J.L. Threlkeld, *Thermal Environmental Engineering*, 3rd ed., Prentice Hall, Upper Saddle River, 1998.

F.C. McQuiston, J.D. Parker, J.D. Spitler, *Heating, Ventilation, and Air Conditioning: Analysis and Design*, 6th ed., Wiley, Hoboken, 2005.

J.W. Mitchell, J.E. Braun, *Principles of Heating, Ventilation, and Air Conditioning in Buildings*, Wiley, Hoboken, 2013.

CHAPTER 3

Psychrometry

"There it is, fog, atmospheric moisture still uncertain in destination, not quite weather and not altogether mood, yet partaking of both."

–Hal Borlamd

- Appreciate what psychrometry is and why it is important in engineering human thermal comfort.
- Comprehend mixture composition quantities including mass fraction, mole fraction, and volume fraction.
- Deduce non-reacting gas mixture properties from the properties of the individual constituents.
- Appreciate ideal gas mixtures, noting the difference with respect to real gas mixtures.
- Apply Dalton's law of additive partial pressures for ideal gas mixtures.
- Employ Amagat's law of additive volumes for ideal gas mixtures.
- Differentiate humidity ratio from relative humidity.
- Note the subtle differences between dew point, adiabatic saturation, and wet-bulb temperatures.
- Appreciate moist air property relationships and approximations.

Nomenclature

c_P Heat capacity at constant pressure; $c_{P,a}$ is constant-pressure heat capacity of dry air, $c_{P,wv}$ is constant-pressure heat capacity of water vapor

E Energy; E_{evap} is the energy of evaporation, E_{in}' is the rate of energy entering a system, E_{kettle} is the energy usage of a kettle, E_{out}' is the rate of energy leaving a system

H Enthalpy; H_a is the enthalpy of dry air, H_{wv} is the enthalpy of water vapor

h Enthalpy; h_a is the specific enthalpy of dry air, h_g is the specific enthalpy of a substance in its gaseous state, h_{sat} is the specific enthalpy at saturation, h_{wv} is the specific enthalpy of water vapor, Δh_a is the change in specific enthalpy of dry air, Δh_{evap} is the enthalpy of evaporation per unit mass

HVAC   Heating, ventilation, and air conditioning

k Number of species

M Molar or molecular mass; M_i is molar mass of species i, $M_{mix} = m_{tot}/n_{tot}$ is the apparent (average) molar (molecular) mass of a mixture

m Mass; m_a is the mass of air, m_i is mass of species i, m_f' is mass flow rate of a substance in liquid phase (fluid), m_{tot} is the total mass, m_{wv} is the mass of water vapor

n Number of moles; n_i is the number of moles of species i

P Pressure; $P_{dry\ air}$ is the pressure of dry air, P_{mix} is the pressure of the mixture, P_{wv} is the pressure of water vapor

R Gas constant; $R_{dry\ air}$ is the gas constant of dry air, $R_{mix} = \Re/M_{mix}$ is the gas constant of the mixture

T Temperature; T_{dp} is the dew point temperature, T_{dry} is the dry-bulb temperature, T_{wet} is the wet-bulb temperature

v Specific volume

x Mass fraction; x_{H_2O} is the mass fraction of H_2O, x_i is mass fraction of species i

y Mole fraction; y_i is mole fraction of species i, y_{wv} is mole fraction of water vapor, $y_{wv,sat}$ is mole fraction of water vapor at saturation

Greek and Other Symbols

η Efficiency; $\eta_{kettle} = E_{evap}/E_{kettle}$ is the efficiency of a kettle

Ξ Degree of saturation

ϕ Relative humidity, $\phi = y_{wv}/y_{wv,sat}$

ω Absolute humidity, or, humidity ratio; ω_{sat} is the humidity ratio at saturation

$\mathfrak{R}$ Universal gas constant

$\forall$ Volume; $\forall_i$ is the volume of species i, $\forall_{mix}$ is the volume of a mixture

3.1 THE SIGNIFICANCE OF MOISTURE IN AIR

Simply put, psychrometry is the study and/or dynamics of moist air. Psychrometrics and hydrometry are synonyms for psychrometry. It is a given that heating and cooling of air are the core of engineering human thermal comfort. After all, we are truly very sensitive beings. Within the scope of engineering a suitable indoor environment for human thermal comfort, it is evident that the moisture content of air has a major effect on the following:

1) *Building Occupants.* Inside a typical building, air that is too moist can foster the growth of various fungi on the occupant's skin. Ringworm is possibly the most notoriously familiar fungus to most of us. On the other hand, eczema, nose bleeding, and asthma can result when the air is too dry. In a setting such as a hospital, the air needs to be maintained slightly dryer than the ideal condition for human thermal comfort to keep germs and viruses in check. Some recent studies, however, found that going on the dry side actually causes various issues [Taylor, 2016]. The 50% relative humidity, which is more or less ideal for human thermal comfort, is also the condition to mitigate all kinds of diseases.

2) *Building Materials.* Dryness is less of a problem than moisture when it comes to building materials. One exception is a museum, where the artwork,[1] which can be part of the building, may be damaged when the air is too dry. Moisture, however, tends to cause the building materials to expand and warp. Mold, some of which is toxic [ASHRAE Journal, 2002], can be a serious issue when the moisture is not properly regulated.

---

[1]The truth is hard to swallow. Real artwork is worth a lot more than human life. Its lofty worth, as expected, requires high maintenance, i.e., within very stringent ranges of temperature and humidity. It is said that Mona Lisa and her equals were escorted out of Paris in air-conditioned vehicles, way ahead of countless targeted living beings, as the Nazis headed for Paris.

From the energy usage perspective, the removal or addition of moisture from or to the air, depending on how this is realized, can be very costly. This is largely because of the high latent heat of water, at approximately 2450 kJ/kg at room temperature.

EXAMPLE 3.1. THE HIGH COST OF HUMIDIFICATION

Given: During winter, Professor Ting needs, other than a scenic window view, approximately 1.5 L of water vapor in his cozy office to be comfortable. He has an old 1-kW electric kettle with a relatively low efficiency. It takes 100 minutes to completely evaporate 1.5 L of water. It happens that the air exchange and heat loss in his office more or less balance in the continual refurnishing of 1.5 L of hot steam every 100 minutes, as long as Professor Ting is laboring for the students in his office.

Find: The amount of energy required to evaporate the 1.5 L of water every 100 minutes. The efficiency of Professor Ting's old electric kettle.

Solution:

The amount of energy required to evaporate 1.5 L of water, assuming the water is initially at the boiling point,

$$E_{evap} = m\Delta h_{evap} = 0.0015\,\text{m}^3 \times 1000\,\text{kg/m}^3 \times 2450\,\text{kJ/kg} = 3.675 \times 10^6\,\text{J}$$

The amount of energy consumed by the 1-kW electric kettle every 100 minutes,

$$E_{kettle} = 1000\,\text{W} \times 100\ \text{min} \times 60\,\text{s/min} = 6 \times 10^6\,\text{J}$$

The efficiency of the kettle,

$$\eta_{kettle} = E_{evap}/E_{kettle} = 0.61,\ \text{or},\ 61\%$$

In this particular case, roughly 39% of the energy is lost but not wasted. The loss is almost completely transformed into thermal heating of the occupied space. This auxiliary heating prevents Professor Ting from having brain freeze due to otherwise insufficient heating by the energy-saving technology. The fancy energy-saving technology utilizes slow-moving supply air, which tends to be inadequately heated by the concrete mass, delivering less-than-thermally-comfortable warm air into Dr. Ting's office.

The purporse here is to highlight the high-energy cost associated with humid-ification. In this case, tens of megajoules of energy per day is required. Fortu-nately, there are substantially less energy-intensive means to humidify a space.

3.2 GAS MIXTURES

In human comfort engineering, the involved air is treated as a mixture of two pure substances, i.e., dry air and water vapor. Accordingly, it is prudent to revisit a mix-ture of pure substances thermodynamically.

3.2.1 A Mixture of Pure Substances

A gas consisting of a mixture of non-reacting species can be treated as a pure sub-stance. The properties of this non-reacting gas mixture is then a function of the constituent gases and the amount of each gas in the mixture. This allows the devel-opment of relevant rules to deduce mixture properties from a knowledge of mixture composition and the properties of the constituent gases.

The mass fraction of a constituent gas,

$$x_i = m_i/m_{tot},\tag{3.1}$$

where m_i is the mass of component i and m_{tot} is the total mass of the mixture, i.e., the sum of the masses of all components. Mixing 18 g of water vapor with 2800 g of nitrogen gas gives 2818 g of nitrogen plus water vapor mixture, see Fig. 3.1. The mass fraction of water vapor in the resulting mixture, x_{H_2O}, is 0.6%.

The mole fraction of component i,

$$y_i = n_i/n_{tot},\tag{3.2}$$

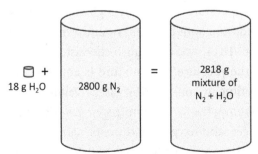

Figure 3.1. Mass fractions of constituent gases in a mixture (created by D. Ting).

where n_i is the number of moles of species i and n_{tot} is the total number of moles of the mixture, i.e., all species together. For the example illustrated in Fig. 3.1, we can view it as 1 mole of water vapor combining with 100 moles of nitrogen gas, resulting in 101 moles of nitrogen–water vapor mixture. The mole fraction of water vapor in the mixture, y_{H_2O}, is 0.99%.

The mass of a substance,

$$m = n\,M, \tag{3.3}$$

where n is the number of moles and M is the molar or molecular mass. For the illustration portrayed in Fig. 3.1, the mass of water vapor is equal to 1 mole times 18 g/mole, which is 18 g.

The apparent (average) molar (molecular) mass of a mixture,

$$M_{mix} = m_{tot}/n_{tot} = m_{mix}/n_{mix} = \sum (n_i M_i)/n_{mix} = \sum_{i=1}^{k} \left(y_i M_i\right), \tag{3.4}$$

where k is the number of species. Similarly, the apparent gas constant of a mixture,

$$R_{mix} = \mathfrak{R}/M_{mix}, \tag{3.5}$$

where $\mathfrak{R}$ is the universal gas constant.

It is clear that the mass and mole fractions of a mixture are related by

$$x_i = m_i/m_{tot} = (n_i M_i)/(n_{mix}M_{mix}) = y_i M_i/M_{mix}. \tag{3.6}$$

3.2.2 Handling Ideal Gas Mixtures

For an ideal gas, the constituent molecules are spaced far apart, and thus, the behavior of a molecule is not influenced by the presence of the other molecules. This is the case when the temperature is relatively high and the pressure is relatively low. Under this ideal condition, both Dalton's law and Amagat's law hold exactly.

John Dalton [1766–1844] was an English chemist, physicist, and meteorologist. He postulated that the pressure of a mixture is equal to the sum of the partial pressures of the constituent, non-reacting gases [Dalton, 2018]. This is depicted in Fig. 3.2, where combining the 0.79 atm nitrogen gas with the 0.21 atm oxygen gas gives 1 atm of nitrogen and oxygen mixture, i.e., air. View it in another way, the pressure in the container would be lowered by 0.21 atm in the absence of oxygen, the far left, nitrogen-only case. Dalton first observed this "pressure additive" phenomenon in 1801 [Dalton, 1802]. It is derivable that this **Additive Partial Pressures** law holds when there are negligible interactions between the constituent gases. This condition applies for ideal gases.

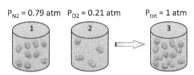

$P_{N2} = 0.79$ atm $P_{O2} = 0.21$ atm $P_{tot} = 1$ atm

Figure 3.2. Dalton's Law of Additive Partial Pressure (created by Z. Yang). The three containers are of identical volume.

In short, Dalton's Law of Additive Partial Pressures states that the pressure of a gas mixture is equal to the sum of the pressures each gas would exert if it existed alone. Accordingly, at the mixture temperature, T, and volume, $\forall$,

$$P_{mix} = \sum_{i=1}^{k} (P_i). \tag{3.7}$$

Amagat's **Law of Additive Volumes** says that, for ideal gases, the volume of the mixture is made of the volumes of each constituent gas when it existed alone at the mixture temperature, T_{mix}, and pressure, P_{mix}. This is portrayed in Fig. 3.3, where,

$$\forall_{mix} = \sum_{i=1}^{k} (\forall_i). \tag{3.8}$$

For the particular case shown, 79 cm³ of nitrogen at temperature, T, and pressure, P, plus 21 cm³ of oxygen at T and P gives 100 cm³ of mixture of nitrogen and oxygen, which is roughly the composition of atmospheric air.

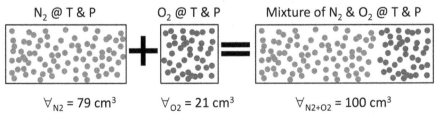

N_2 @ T & P O_2 @ T & P   Mixture of N_2 & O_2 @ T & P

$\forall_{N2} = 79$ cm³ $\forall_{O2} = 21$ cm³ $\forall_{N2+O2} = 100$ cm³

Figure 3.3. Amagat's Law of Additive Volumes (created by Z. Yang). The temperature and pressure inside all three containers are the same.

Moving forward, we will not look at separating oxygen and nitrogen. Rather, the roughly 79% nitrogen and approximately 21% oxygen will be treated as a single constituent, i.e., dry air. For heating, ventilation, and air conditioning (HVAC) applications, it is both necessary and convenient to treat the water vapor as a unique constituent. In other words, air consists of dry air plus water vapor. The amount of water vapor, though small in quantity, varies rather extensively in heating and cooling processes.

3.2.3 Dealing with Atmospheric Air

Atmospheric air consists of about 78% nitrogen, 21% oxygen, along with traces of argon, carbon dioxide and other species, and water vapor, i.e.,

$$\text{air} \approx 78\% \, N_2, \, 21\% \, O_2, + \text{traces of Ar, } CO_2, \text{ etc.} + \text{water vapor.} \qquad (3.9)$$

In HVAC applications, the role played by argon and other minute species is inconsequential and, thus, can be neglected.[2] The amounts of nitrogen, oxygen, as well as argon are largely fixed, but not water vapor. It is interesting to note that the molecular weight of dry air is about 29 kg/kmol, while that of water vapor is around 18 kg/kmol (see Table 3.1). Surprise, surprise, water vapor is lighter than air!

Table 3.1. Water vapor versus dry air in atmospheric air.

Air =	dry air +	water vapor
Composition	78% N_2 + 21% O_2	<1% H_2O
Molecular weight [kg/kmol]	28.966	18.016 (not heavier!)

The air that we deal with in heating and cooling for thermal comfort is always moist, i.e., it always contains some water vapor. The amount of water vapor, however, tends to change significantly. When the weather is hot, the air tends to be humid,[3] and thus, cooling the humid air leads to the removal of water vapor, i.e., via condensation. In nature, this cooling and condensation process is most beautifully portrayed as morning dew on leaves. During the cold season, on the other

---

[2]Obvious exceptions are the harmful pollutants, which may occur only in parts per millions, but can cause serious health problems. This is dealt with under indoor air quality.

[3]An obvious exception is the arid regions. Even in these regions, however, creatures such as the Australian thorny devils can condense water vapor out of the "thin and dry" desert air to quench their thirst.

hand, even relatively humid ambient air when heated becomes too dry for thermal comfort. Consequently, moisturization (humidification) is necessary.

For an ideal gas,

$$P\forall = m\,(\mathfrak{R}/M)\,T, \qquad (3.10)$$

or,

$$Pv = RT, \qquad (3.11)$$

where the universal gas constant, $\mathfrak{R}$ = 8.314 kJ/kmol·K, and v is the specific volume, i.e., v = $\forall$/m. For dry air, the gas constant R = 8.314 kJ/kmol·K/28.966 kg/kmol = 287 J/kg·K. For the lighter water vapor, R = 8.314 kJ/kmol·K/18.016 kg/kmol = 463 J/kg·K. Concerning the molecular weight, M, we recall that Avogadro's law states that 1 mole of any substance (gas species) contains 6.023 × 10^23 molecules.

In heating and cooling load calculations, invoking the ideal gas assumption results in less than 1% error [Threlkeld, 1970]. This is well within the acceptable uncertainties associated with engineering practice. Furthermore, the dry air can be treated as an ideal gas with a constant heat capacity, c_p = 1.005 kJ/kg·K (0.240 Btu/lbm·R). As such, the enthalpy (thermal energy) change associated with a change in the dry air temperature of ΔT,

$$\Delta h_a = c_p \Delta T = 1.005\,\text{kJ/kg} \cdot \text{K}\,\Delta T, \qquad (3.12)$$

where ΔT is in °C or K. Typical changes in c_p at atmospheric pressure over the range of temperature encountered in HVAC are given in Table 3.2. It is clear that for the engineering applications involved, setting the c_p of dry air to be equal to 1.005 kJ/kg·K is a sound assumption.

Table 3.2 Typical variation of dry air c_p with temperature.

T [°C]	−10	0	10	20	30	40	50
c_p [kJ/kg·K]	1.0038	1.0041	1.0045	1.0049	1.0054	1.0059	1.0065

Also, for an ideal gas, the enthalpy is only a function of temperature, i.e., h = f(T). For temperatures less than roughly 50°C, the constant-h lines coincide with the constant-T lines, see Fig. 3.4. To that end, the enthalpy of the water vapor is approximately equal to that of dry air, i.e.,

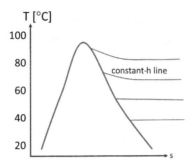

Figure 3.4. The constant-h lines coincide with the constant-T lines for an ideal gas (created by D. Ting).

$$h_{wv} \cong h_{dry\ air}\ (T) \cong 2500.9 + 1.82\ T, \tag{3.13}$$

where T is in °C and h is in kJ/kg.

3.3 HUMIDITY RATIO VERSUS RELATIVE HUMIDITY

The word "humidity" has been taken for granted to imply relative humidity. Relative humidity measures the amount of water vapor in the air with respect to the maximum amount of water vapor the air at that thermodynamic state (given temperature and pressure) can hold. Simply put, the **relative humidity**,

$$\varphi = y_{wv}/y_{wv,sat}, \tag{3.14}$$

where y_{wv} is the mole fraction of water vapor in air (dry air and water vapor mixture), $y_{wv,\ sat}$ is the mole fraction of water vapor at saturation at the same temperature and pressure. Recall that saturation refers to the condition where air contains the maximum amount of water vapor, and any addition of water vapor into the system will cause some water vapor to condense into liquid water.

Invoking the ideal gas principle, or Dalton's law of additive pressures, the total pressure,

$$P_{tot} = P_{dry\ air} + P_{wv}, \tag{3.15}$$

where $P_{dry\ air}$ is the partial pressure of the air without any water vapor, and P_{wv} is the partial pressure induced by the water vapor. In terms of partial pressure of water vapor, saturation corresponds to the maximum partial pressure that the water vapor can contribute as far as the total pressure is concerned. It follows that we can express the relative humidity in terms of pressures,

$$\varphi = P_{wv}/P_{wv,sat}, \tag{3.16}$$

where P_{sat} is the saturation pressure of water vapor. It is clear that increasing the mole fraction of water vapor, adding water vapor, increases the partial pressure of water vapor, and thus, the relative humidity, until saturation.

The humidity ratio is also called the absolute or specific humidity. As the name implies, **humidity ratio** is the ratio of the mass of water vapor with respect to the mass of (dry) air, i.e.,

$$\omega = m_{wv}/m_{dry\,air}. \tag{3.17}$$

Invoking the ideal gas assumption, this can be expressed as

$$\omega = \left(P_{wv}\forall/R_{wv}T\right)/\left(P_{dryair}\forall/R_{dryair}T\right) = \left(P_{wv}/R_{wv}\right)/\left(P_{dryair}/R_{dryair}\right) \tag{3.18}$$
$$= 0.622\,P_{wv}/P_{dryair},$$

We see that 0.622 is the water vapor/dry air gas constant ratio. As the total pressure is the sum of the water vapor partial pressure and dry air partial pressure, the humidity ratio may also be expressed as

$$\omega = 0.622\,P_{wv}/\left(P_{tot} - P_{wv}\right). \tag{3.19}$$

A less-known humidity parameter is the degree of saturation. The **degree of saturation** is simply the humidity ratio with respect to the humidity ratio at saturation, i.e.,

$$\Xi = \omega/\omega_{sat}. \tag{3.20}$$

A summary of the three measures of moisture in the air is given in Table 3.3. The relationship between relative humidity and humidity ratio derived based on the ideal gas assumption can be applied in typical HVAC applications for all practical purposes. Note that neither relative humidity nor the degree of saturation is defined when the temperature of moist air exceeds the saturation temperature of pure water corresponding to the moist air pressure. Fog is composed of tiny (liquid) water droplets suspended in air. Therefore, in the presence of fog, the relative humidity is, in reality, greater than 100%. The amount of the tiny water droplets suspended in air is accounted as part of the mass of water vapor, m_{wv}, which is used to define the humidity ratio, $\omega = m_{wv}/m_{dry\,air}$. Hence, the wetness (moisture content) of air in this breath-taking scenario can be described (characterized) by the humidity ratio, along with the pressure and temperature. Also interesting to note is that at atmospheric pressure, the humidity ratio at saturation, ω_{sat}, approaches infinity at 100°C.

Table 3.3. Relative humidity, humidity ratio, and degree of saturation.

Relative humidity	Humidity ratio (absolute humidity)	Degree of saturation
$\phi = P_{wv}/P_{wv,\,sat}$	$\omega = m_{wv}/m_{dry\;air}$	$\Xi = \omega/\omega_{sat}$
For ideal gases, $\phi = (1 + 0.622/\omega_{sat})/(1 + 0.622/\omega)$		

EXAMPLE 3.2. COOLING THE AIR INSIDE A WELL-SEALED SPACE

Given: The air inside a well-sealed space at 27°C and 50% is cooled to 25°C.
  Find: What can you say about (a) the relative humidity, (b) the humidity ratio, and (c) the wet-bulb temperature?

Solution:

(a) The relative humidity increases with the drop in temperature.
(b) Cooling 27°C and 50% relative humidity air to 25°C does not lead to condensation. This sensible cooling can easily be seen on a psychrometric chart, covered in chapter 4. Therefore, humidity ratio remains unchanged.
(c) The wet-bulb temperature decreases. Again, this process is clearly illustrated on a psychrometric chart.

3.4 DRY-BULB, DEW POINT, ADIABATIC SATURATION, AND WET-BULB TEMPERATURES

The **dry-bulb temperature**, T_{dry}, is the temperature of moist air measured by a perfectly dry sensor. It is the true temperature of moist air at rest. In short, the dry-bulb temperature is the thermodynamic temperature of the air. In our everyday conversation, the pleasantness of the weather is customarily described by the temperature, which is the dry-bulb temperature, and the humidity, which implicitly denotes the relative humidity.

The **dew point temperature** is the temperature at which condensation begins when the (moist) air is cooled at a constant pressure. An everyday example is a bottle of cold water sitting in a room, especially during the summer time, see Fig. 3.5, where droplets of condensate transpire. One could use a line of cold water bottles

Figure 3.5. Condensation of water vapor on a cold water bottle (photo taken by S. K. Mohanakrishnan).

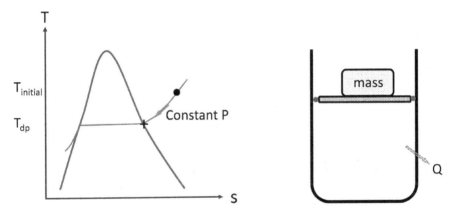

Figure 3.6. Cooling of air at constant pressure to its dew point temperature (created by D. Ting). This can be achieved by removing heat from a container of moist air with a freely moving piston until water condenses inside the container.

at a series of temperatures to gauge the humidity in the room. The warmest temperature where condensation just barely takes place is the dew point temperature. This cooling of air to its dew point temperature is illustrated on the T–s diagram in Fig. 3.6. The constant-pressure cooling is delineated by the container of (moist) air with a freely moving piston. Using the cold water bottle example, the room air next to the cold surface of the bottle when cooled moves along the constant pressure line. When it reaches the dew point temperature, T_{dp}, the first condensate appears. Morning dew, which nourishes the meadow eventuates when humid atmospheric air, is cooled to its dew point temperature.

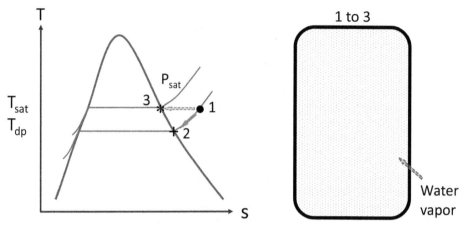

Figure 3.7. Dew point temperature versus saturation temperature (created by D. Ting). By adding water vapor into a constant-volume container, the pressure of the container is increased accordingly. When there is too much water vapor for the given amount of air to retain, the first condensate forms, and this is the saturation temperature.

The dew point temperature can also be viewed as the temperature at which moist air at a given humidity ratio becomes saturated. In Fig. 3.7, Process 1–2 describes the constant pressure, at P_{wet}, cooling of superheated water vapor at the dry-bulb temperature, T_{dry}, to a saturated-vapor state at the dew point temperature, T_{dp}. This can be physically achieved by removing heat from the moist air inside the cylinder sealed with a freely moving piston, which maintains a constant in-cylinder pressure. Another way to cause condensation is by fixing the position of the lid, i.e., keeping the volume of the cylinder fixed. Adding water vapor into the cylinder causes the pressure to rise, Process 1–3. Some water vapor condenses into liquid water at Point 3, where the temperature is called the **saturation temperature**. Note that some heat needs to be removed to compensate for the slight increase in temperature caused by the small rise in pressure from P_{wet} to P_{sat} associated with the addition of water vapor into the constant-volume cylinder.

It is thus clear that, thermodynamically, the humidity is a function of the vapor pressure and the dew point temperature. Although theoretically sound, measuring the vapor pressure is not rational practically. Therefore, a more practical deduction of humidity, based on temperature measurements, is sought. Figure 3.8 depicts an ideal case concerning the **adiabatic saturation temperature**. As the unsaturated air passes through the long adiabatic channel containing a pool of water, some liquid water evaporates into water vapor, adding to the existing moisture of the flow stream. With the increase in the moisture content, the air temperature decreases.

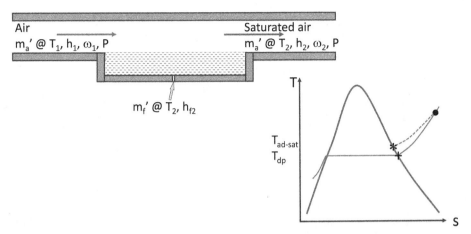

Figure 3.8.   Adiabatic saturation temperature (created by D. Ting). The channel and the pool of water are adequately long so that the air reaches saturation.

This is because some energy in the air is used to evaporate the liquid water into water vapor, i.e., the latent heat of vaporization where the attraction between water molecules is broken, letting the individual water molecules move freely in the gaseous phase. If the channel is long enough, the air will exit the channel as saturated air at T_2, which is equal to $T_{ad\text{-}sat}$, the adiabatic saturation temperature. This process is also illustrated on the T–s diagram. It is worth stressing that the partial pressure of the evaporated water vapor raises the air pressure slightly, i.e., from P_1 to P_2.

Let us apply the conservation laws for the steady-flow adiabatic saturation process depicted in Fig. 3.8. "Dry air in" is equal to "dry air out," i.e., the conservation of dry air leads to

$$m_{a1}' = m_{a2}' = m_a',$$

(3.21)

where apostrophe signifies rate. Similarly, the conservation of water vapor gives

$$m_{w1}' + m_f' = m_{w2}',$$

(3.22)

which says that the existing water vapor in the incoming air plus that evaporated from the pool of water is equal to the amount of water vapor in the outgoing air, which is saturated in the ideal case with a long channel. This can be rewritten as

$$m_a'\omega_1 + m_f' = m_a'\omega_2,$$

(3.23)

where ω_2 is ω_{sat} in the ideal case. The evaporation rate can be expressed in terms of the increase in the humidity ratio from inlet to outlet, that is

$$m_f' = m_a'(\omega_2 - \omega_1). \tag{3.24}$$

This says that the mass of water evaporated is equal to the moisture in the exiting air minus that in the incoming air.

According to the first law of thermodynamics, energy cannot be created nor destroyed. Invoking this energy conservation principle to the system of concern gives

$$E_{in}' = E_{out}', \tag{3.25}$$

i.e., the rate of energy entering the system is equal to that exiting the system under the steady-state condition. Considering the flow of energy from the incoming air and the pool of water into the outgoing air, we have

$$m_a' h_1 + m_f' h_{f2} = m_a' h_2. \tag{3.26}$$

This expression says that the enthalpy of the incoming moist air plus that from evaporation is equal to that of the outgoing moist air. Section 3.5 will disclose the background behind enthalpy, heat capacity, etc., when dealing with moist air. Dividing Eq. 3.26 by the mass flow rate of dry air results in

$$h_1 + (\omega_2 - \omega_1) h_{f2} = h_2. \tag{3.27}$$

This, as explained in Section 3.5, can also be expressed in terms of heat capacity and temperature, i.e.,

$$\left(c_P T_1 + \omega_1 h_{g1}\right) + (\omega_2 - \omega_1) h_{f2} = \left(c_P T_2 + \omega_2 h_{g2}\right). \tag{3.28}$$

The pressure is by-and-large constant, other than a tiny increase caused by the very small P_{wv} added. Consequently, the enthalpy at saturation coming out at the end of the channel, h_{sat2}, and h_{f2} are solely dependent on T_2, and T_2 is a function of h_1, ω_1, and P. As such, T_2 is a thermodynamic property of the incoming air at State 1: we call this property the thermodynamic wet-bulb temperature, T_{wet}. Thus, Eqs. 3.27 and 3.28 can be re-expressed as

$$h + (\omega_{wet} - \omega) h_{f2} = h_{sat}. \tag{3.29}$$

$$\left(c_P T + \omega h_g\right) + (\omega_{wet} - \omega) h_{f2} = (c_P T_{wet} + \omega_{wet} h_{sat}). \tag{3.30}$$

In other words, for given values of h, ω, and P, i.e., a given moist air state, the thermodynamic wet-bulb temperature, T_{wet}, is that value that satisfies this equation. Physically, the incoming air is humidified to saturation before exiting. As such, T_2 is equal to T_{wet}, i.e., the dry-bulb temperature is equal to the wet-bulb temperature at saturation.

Regarding the dry-bulb and wet-bulb temperatures, it has been shown that the difference is zero for saturated air. In HVAC practice, the rule of thumb is that there is a 5.5°C difference between the dew point temperature and the dry-bulb temperature at a relative humidity of 70%. For dryer air at a relative humidity of 50%, there is more room for evaporation, further lowering the wet-bulb temperature, leading to a ($T_{dry} - T_{wet}$) temperature difference of about 11°C, at atmospheric pressure.

A more practical way, instead of using a very long channel with a pool of water at the bottom as shown in Fig. 3.8, to measure the wet-bulb temperature is via a thermometer with a wetted wick over which air flows at a specified velocity between 2.5 and 5 m/s [Threlkeld, 1970], see Fig. 3.9. Alternate to having a fan supplying the needed flow to appropriately evaporate the water from the wick, a sling psychrometer operates based on moving the wet bulb with respect to the air. At equilibrium, the wet thermometer bulb is at a temperature, T_{wet}, at which the rate of evaporation times the heat of vaporization is equal to the heat transferred to the bulb; see Fig. 3.10. As the air next to the wet bulb loses energy, its temperature

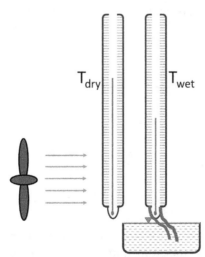

Figure 3.9.   Psychrometer for deducing air humidity from the dry-bulb and wet-bulb temperatures (created by D. Ting). Some minimum air movement is required to convect the highly moist air away from the wet bulb.

drops. As mentioned before, it is apparent that the dryer the air, the faster the evaporation, and thus, the lower the wet-bulb with respect to the dry-bulb temperature.

An indirect, but practical, way to approximate the humidity of air is via a hydrometer such as that shown in Fig. 3.11. Materials such as a human hair elongate as the humidity increases. This kind of hydrometer, however, needs to be properly calibrated against a standard before usage. It is also not meant for providing a highly accurate measurement.

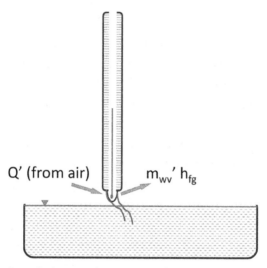

Figure 3.10. The heat balance of a wet bulb (created by D. Ting). At equilibrium, the amount of heat removed from the wet bulb is equal to that convected into it.

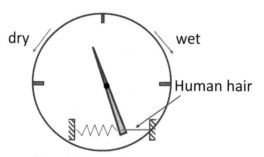

Figure 3.11. A hydrometer for estimating air humidity (created by D. Ting). Human hair expands with moisture, and, hence, can be calibrated to indicate the humidity in the air.

3.5 MOIST AIR PROPERTIES AND PROPERTY RELATIONSHIPS AND APPROXIMATIONS

The enthalpy of (moist) air consists of the enthalpy of the dry air and that of the water vapor, i.e.,

$$H = H_a + H_{wv} = m_a\,h_a + m_{wv}\,h_{wv}. \tag{3.31}$$

The specific enthalpy of moist air, h, is defined as $h \equiv H/m_a$, i.e., per unit mass of **dry** air. On that, we can divide Eq. 3.31 by m_a to give

$$h = h_a + (m_{wv}/m_a)\,h_{wv} = h_a + \omega\,h_{wv}. \tag{3.32}$$

Note that h_{wv} is the amount of enthalpy of water vapor per unit mass of water vapor, m_{wv}. Hence, the term $(m_{wv}/m_a)\,h_{wv}$ or $\omega\,h_{wv}$ is the enthalpy of the water vapor per unit mass of dry air. In short, every term in Eq. 3.32 depicts the amount of enthalpy per unit mass of dry air. The use of unit mass of dry air as the base conveniently eliminates any complication associated with the ever-changing mass of water vapor along an air duct.

For a perfect gas, h = f(T) only, i.e.,

$$h_a = \int c_{P,a}\,dT \tag{3.33}$$

and

$$h_{wv} = \int c_{P,wv}\,dT. \tag{3.34}$$

Assuming constant specific heat capacities, the above can be simplified into

$$h_a = c_{P,a} \int dT \tag{3.35}$$

and

$$h_{wv} = c_{P,wv} \int dT. \tag{3.36}$$

Selecting the reference state $h_a = 0$ at T = 0°C (or °F for English units) leads to

$$h_a = c_{P,a}T - c_{P,a}(0) = c_{P,a}\,T \tag{3.37}$$

and

$$h_{wv} = c_{P,wv} T + h_{sat}, \tag{3.38}$$

where h_{sat} is the enthalpy of saturated water vapor at 0°C. This is not necessarily zero because we already chose the enthalpy of the dry air, h_a, to be zero at 0°C. Consequently, h_{sat} with respect to h_a at 0°C is non-zero.

The total enthalpy of moist air,

$$h = h_a + \omega\, h_{wv} = c_{Pa}\, T + \omega\, h_{wv} = c_{Pa}\, T + \omega\, (c_{Pwv}\, T + h_{sat}). \tag{3.39}$$

However,

$$c_P = c_{Pa} + \omega\, c_{Pwv}, \tag{3.40}$$

upon which, we have

$$c_P T = c_{Pa} T + \omega\, c_{Pwv} T. \tag{3.41}$$

Namely, the enthalpy of moist air per unit mass of dry air,

$$h = c_P T + \omega\, h_{sat}. \tag{3.42}$$

The total entropy of moist air,

$$S = S_{ao} + \Delta S_a + \Delta S_{a\,mix} + S_{wvo} + \Delta S_{wv} + \Delta S_{wv,mix}, \tag{3.43}$$

where S_{ao} is the dry air entropy at a reference state, ΔS_a is the change in dry air enthalpy from the reference state, $\Delta S_{a,mix}$ and $\Delta S_{wv,mix}$ are the mixing entropies for the dry air and water vapor, respectively, S_{wvo} is the water vapor entropy at a reference state, and ΔS_{wv} is the change in water vapor enthalpy from the reference state. These, in a sense, account for the fact that partial pressures, P_a and P_{wv}, and not the total pressure, P, are needed in evaluating the entropies of the components. After some manipulations, the specific entropy can be expressed as

$$s = (c_{Pa} + c_{Pwv}\, \omega)\, \ln{(T/T_o)} - R_a\, \ln{(P_a/P_o)} + \omega\, [s_{wvo} - R_{wv}\, \ln{(P_{wv}/P_o)}]. \tag{3.44}$$

In many practical cases, various approximations may be invoked to ease hand calculations. For a process from State 1 to State 2, the specific enthalpy change,

$$\Delta h = h_2 - h_1 = (c_{Pa}\, T_2 + \omega_2\, c_{Pwv}\, T_2 + \omega_2\, h_{sat}) - (c_{Pa} T_1 + \omega_1\, c_{Pwv}\, T_1 + \omega_1\, h_{sat}). \tag{3.45}$$

The average temperature, $T_{avg} = (T_1 + T_2)/2$, and the temperature change, $\Delta T = T_2 - T_1$. Similarly, the average humidity ration, $\omega_{avg} = (\omega_1 + \omega_2)/2$, where the humidity rise, $\Delta\omega = \omega_2 - \omega_1$. With these, Eq. 3.45 can be rewritten as

$$\Delta h = \left(c_{Pa} + \omega_{avg}\, c_{Pwv}\right)\Delta T + \left(c_{Pwv}\, T_{avg} + h_{sat}\right)\Delta\omega. \tag{3.46}$$

The two contributors to the enthalpy change are sensible change,

$$\Delta h_S = \left(c_{Pa} + \omega_{avg}\, c_{Pwv}\right)\Delta T, \tag{3.47}$$

and latent change,

$$\Delta h_L = \left(c_{Pwv}\, T_{avg} + h_{sat}\right)\Delta\omega. \tag{3.48}$$

It is clear that sensible change corresponds to a variation in the (dry-bulb) temperature, ΔT, and latent change is associated with an alteration in the humidity ratio, $\Delta\omega$. Note that

$$c_{Pa} + \omega_{avg}\, c_{Pwv} = c_{P,avg}, \tag{3.49}$$

and

$$c_{Pwv}\, T_{avg} + h_{sat} = h_{avg}. \tag{3.50}$$

Therefore,

$$\Delta h_S = c_{P,\,avg}\,\Delta T, \tag{3.51}$$

and

$$\Delta h_L = h_{avg}\,\Delta\omega. \tag{3.52}$$

As the numerical values of $c_{P,avg}$ and h_{avg} are relatively insensitive to the humidity ratios and temperatures, respectively, that commonly occur in HVAC processes, they can be assumed to be constants. In other words,

$$c_{P,\,avg} \approx 1.02\ \text{kJ/kg}_a \cdot \text{°C}\ (0.245\ \text{Btu/lbm}_a \cdot \text{°F}), \tag{3.53}$$

or

$$h_{avg} \approx 2500 \sim 2700\ \text{kJ/kg}_w\ (1050 \sim 1150\ \text{Btu/lbm}_w). \tag{3.54}$$

With software such as EES (Engineering Equation Solver) readily available, these approximations are seldom invoked. Nevertheless, a good engineer should always execute sample hand calculations on the back of an envelope to make sure the computations involving software are reliable.

PROBLEMS

Problem 3.1

What is the molecular weight of dry air composed of the following species?

Species	Volumetric Composition	Molecular Mass [kg/kmol]
O_2	0.2095	32
N_2	0.7809	28.02
Ar	0.00933	39.94
CO_2	0.003	44.01
Neon	0.000018	20.18
He	0.000005	4
H_2	0.000005	2.02
Krypton	0.000001	83.8
Xenon	9×10^{-8}	131.29

If there is 5% volumetric composition of water vapor in the atmospheric dry air (tabulated) and water vapor mixture at 20°C, what is the corresponding molecular weight of the moist air? What are the relative humidity and the humidity ratio?

Problem 3.2

The air inside a well-sealed space is initially at T_{dry} of 21°C and T_{wet} of 9°C. Liquid water at 21°C is used to humidify the air to 50% relative humidity (cool mist humidification). What would happen to T_{dry} and T_{wet}?

Problem 3.3

For 1 kg of atmospheric dry air at 30°C, what is the maximum (at saturation) amount of water vapor? If there is 0.001 kg of water vapor per kg of dry air, what are the humidity ratio, the relative humidity, and the partial pressure of water vapor?

Problem 3.4

Where does the 0.622 of the humidity ratio expression, $\omega = 0.622\, P_{wv}/(P_{tot} - P_{wv})$ come from? What is the main assumption behind this humidity ratio expression?

Problem 3.5

Moist air flows through a long channel as shown in Fig. 3.8. Clearly derive (a) the conservation of dry air, (b) the conservation of moisture, and (c) the conservation of energy.

Problem 3.6

Deduce the dew point temperature at 0.8 atm, 1 atm, and 1.2 atm using the thermodynamics that you have learned, e.g., T–s diagram. For the same relative humidity air, what can you say about having dew formation at higher altitude versus sea level? Why?

Problem 3.7

How much error is introduced if one assumes the enthalpy change of dry air as the energy required to raise 1 kg of saturated air from 15 to 35°C?

REFERENCES

S. Armstrong, "The fundamentals of fungi," ASHRAE Journal, 44(11):18–24, 2002.

J. Dalton, "Essay IV. On the expansion of elastic fluids by heat," Memoirs of the Literary and Philosophical Society of Manchester, 5–2: 595–602, 1082.

J. Dalton, "Daltons-law," 2018, https://www.britannica.com/science/Daltons-law, accessed on August 9, 2018.

S. Taylor, "Would ASHRAE 170 benefit from more health data research?" Engineered Systems, 33(11): 36–41, 2016.

J.L. Threlkeld, *Thermal Environmental Engineering*, Prentice-Hall, Upper Saddle River, 1970.

CHAPTER 4

Psychrometric Chart and Air-Conditioning Processes

"People don't realize that water in the liquid state is very rare in the universe. Away from earth it is usually gas. This moisture is a blessed treasure, and it is our basic duty, if we don't want to commit suicide, to preserve it."

–Jacques-Yves Cousteau.

CHAPTER OBJECTIVES

- Understand the basic construction of the psychrometric chart.
- Appreciate the key fundamental psychrometric processes.
- Become familiar with the utilization of the psychrometric chart.
- Recognize single-zone and multi-zone space conditioning.

Nomenclature

b Bypass factor

c_p Heat capacity at constant pressure; c_{p_a} is the constant-pressure heat capacity of dry air, $c_{p_{wv}}$ is the constant-pressure heat capacity of water vapor

e_{sat} Saturation effectiveness

h Specific enthalpy; h_a is the specific enthalpy of dry air, h_{wv} is the specific enthalpy of water vapor

HVAC   Heating, ventilation, and air conditioning

(Continued)

KE   Kinetic energy; ΔKE is the change in kinetic energy

m   Mass; m' is the mass flow rate, m_a' is the mass flow rate of dry air, m_{wv}' is the mass flow rate of water vapor

P   Pressure

PE   Potential energy; ΔPE is the change in potential energy

Q   Heat; Q_L' is the latent heat transfer rate, Q_S' is the sensible heat transfer rate, Q_{tot}' is the total heat transfer rate

SHR   Sensible heat ratio

T   Temperature; T_{dry} is the dry-bulb temperature, T_{wet} is the wet-bulb temperature

v   Specific volume; $v_{std} = 0.830$ m³/kg (13.33 ft³/lbm), is the specific volume of air at the standard conditions

Greek and Other Symbols

η Efficiency, η_{EC} is the efficiency of an evaporative cooler

ρ Density; for air, $\rho = 1.204$ kg/m³ (0.075 lbm/ft³), at the standard conditions

ω Absolute humidity, or, humidity ratio

4.1 PSYCHROMETRIC CHART CONSTRUCTION

At this point, we can solve heating, ventilation, and air conditioning (HVAC) problems concerning moist air by using the various thermodynamic properties of moist air and the equations relating them. These hand calculations, however, can be very tedious. Utilization of the many available software programs can drastically ease the process. For this reason, routine and/or heavy-duty calculations performed in the HVAC industry are universally executed using software. Frequently, a company would adopt a particular commercial or in-house software, depending on its tradition and/or familiarity with the software. To understand the happenings behind these computer programs, psychrometric charts are extremely useful. On their own, psychrometric charts facilitate quick, though not as speedy as using software on a fast computer, and relatively accurate calculations of most HVAC processes. In the author's narrow and biased view, all good HVAC software should reveal the underlying processes on a psychrometric chart whenever appropriate. No one should trust complex computations based on software performed by someone who cannot carry out a basic calculation using the psychrometric chart.

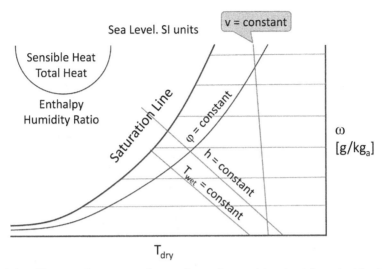

Figure 4.1.   The most common form of pyschrometric chart (created by D. Ting). The horizontal axis is the dry-bulb temperature, and the vertical axis on the far right is the humidity ratio

Recall that the state of air at a given pressure is completely specified by two independent intensive[1] properties. Thus, the other properties can be deduced via the thermodynamic relationships expounded earlier. For any specific pressure (and a range of temperatures), we can generate psychrometric charts. It is obvious that the most common ones are for atmospheric pressure. There are, however, many forms of psychrometric charts depending on the adopted coordinates. The most prevailing choice is to use dry-bulb temperature and humidity ratio as the basic coordinates [Mollier, 1923]. Furthermore, plotting enthalpy, h, as an oblique coordinate, and the humidity ratio, ω, as a rectangular coordinate seems to work best. Along these lines, the more-or-less standard format is shown in Fig. 4.1. It is worth noting that there are many variants within this framework, but these variations are relatively small.

The prevailing types of air-conditioning process trends are highlighted in Fig. 4.2. The most obvious happening when moving to the right is the increase in the dry-bulb temperature. Therefore, the process to the right is heating, and

---

[1]An intensive property is one that "size does not matter." In other words, an intensive property is a physical property that does not depend on the size of the system or the amount of material in the system. In context, temperature, pressure, relative humidity, and absolute humidity are intensive properties.

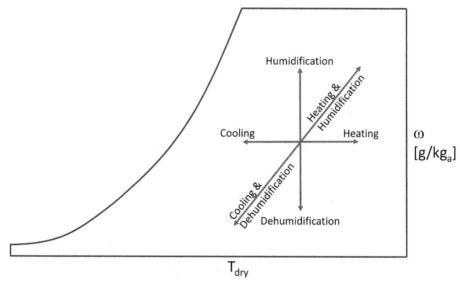

Figure 4.2. The general air-conditioning processes on a pyschrometric chart (created by D. Ting). The four "orthogonal" processes are: (1) sensible heating, horizontally to the right, (2) sensible cooling, horizontally to the left, (3) humidification, vertically up, and (4) dehumidification, vertically down.

to the left, cooling. If these movements are wholly horizontal, then the humidity ratio remains unchanged, and thus, sensible heating takes place when moving to the right, and sensible cooling to the left. Note that sensible heating and cooling, unlike latent heating and cooling, do not involve phase change, i.e., phase change of H_2O. They are related to the macroscopic kinetic energy of the gas molecules, i.e., adding heat increases their kinetic energy and hence, the air temperature. Moving vertically up and down the standard psychrometric chart corresponds most sensitively to changes in the humidity ratio; recall that the vertical axis on the far right corresponds to humidity ratio. As such, humidification of the air takes place when progressing vertically up, and dehumidification takes place when moving vertically down. The other commonly occurring practical process is heating and humidification, moving vertically upward and to the right along a positive slope on the graph. The most common example is leaving a pot of water boiling in a kitchen, raising the kitchen temperature along with the humidity. Opposite to this, moving downward and to the left, is cooling and dehumidification. In real practice, the dehumidification associated with cooling hot and humid air below the saturation temperature does not take place along a straight line as depicted in the figure. The air is actually first sensibly cooled until it reaches the saturation line, from which it moves along the saturation line as it loses moisture. This process will be expounded in the next section.

Other notable features concerning the standard psychrometric chart also worth highlighting include the following:

1. The horizontal axis is the dry-bulb temperature, T_{dry}, but the constant T_{dry} lines are not exactly perpendicular to the horizontal axis, i.e., they slant slightly to the left.

2. The constant T_{wet} lines are not perfectly parallel to the constant h lines. The T_{wet} lines are marginally steeper.

3. The constant specific volume lines are approximately linear and are markedly steeper than the constant T_{wet} lines.

4. In general, the constant relative humidity curves are relatively darker and/or bolder. The 100% relative humidity curve corresponds to the saturation curve.

5. Above (and to the left of) the saturation curved line, the region corresponds to fog conditions. A fog is a mechanical mixture of saturated moist air and water droplets, both at the same temperature.

Dividing a heating or cooling process into two parts, sensible and latent, is a convenient and revealing way to look at the air-conditioning process. In practice, some ratio of these may either be known or required. Therefore, a protractor associated with this, sensible-total heat ratio, is typically given at the top left corner of the psychrometric chart. This protractor provides the scales from which straight lines of constant sensible-latent heat ratio can be extended into the main body of the chart. As the alternate enthalpy-moisture ratio may be sought or given, the protractor also furnishes the scales for constant enthalpy-humidity ratio. These are typically marked on the outside of the protractor, with inner tick marks corresponding to sensible-total heat ratio values.

4.2 PSYCHROMETRIC PROCESSES

Literally all air-conditioning processes can be broken down into a handful of elementary processes. A good understanding of these fundamental processes can drastically facilitate our understanding of the entire heating and cooling requirements, especially in a complex, large-scale, practical setting. For a control volume encompassing a junction of the air-conditioning process line shown in Fig. 4.3, e.g., the first law of thermodynamics for a constant pressure process with

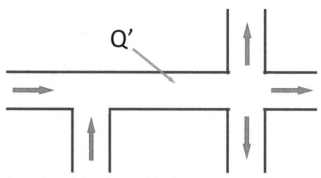

Figure 4.3.   A section of the air-conditioning process line (created by D. Ting).

negligible changes in potential energy, kinetic energy, and no work in or out of the system gives

$$\sum_{in} m'h + Q' = \sum_{out} m'h. \tag{4.1}$$

This states that the sum of all energy associated with the incoming streams per unit time plus the heat input rate is equal to the sum of all outgoing energy per unit time. Let us invoke this, along with the conservation of dry air and of moisture, to explain the handful of fundamental air-conditioning processes.

4.2.1 Sensible Heating and Cooling

Sensible heating and cooling are the simplest air-conditioning processes that involve sensible heat transfer, without the complication associated with moisture addition or removal. As such, the process proceeds horizontally on the standard psychrometric chart. Figure 4.4 presents the straightforward sensible heating of air in an air duct, where the constant-ω process is shown as a horizontal line toward the right on the psychrometric chart. For the corresponding sensible cooling process, the arrow reverses, in the decreasing dry-bulb temperature direction. The sensible cooling portion in practical air conditioning stops a little ahead of the saturation curve, i.e., just before the bulk air temperature reaches the dew point temperature, beyond (below) which some water vapor in the air condenses into liquid water. The practical "little ahead" depends, among other factors, on the level of inhomogeneity in the cooling process, and thus, air temperature. The air next to the cooling coil tends to be cooler, and thus, condenses before the bulk or average air stream drops below the dew point temperature.

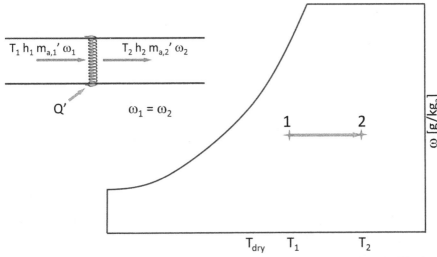

Figure 4.4. Sensible (constant-ω) heating and cooling (created by D. Ting).

The conservation of the mass of dry air mass is simply

$$m_{a,1}' = m_{a,2}' = m_a'. \tag{4.2}$$

Similarly, moisture in is equal to moisture out, i.e.,

$$m_{a,1}'\omega_1 = m_{a,2}'\omega_2. \tag{4.3}$$

Substituting Eq. 4.2 into Eq. 4.3 gives

$$\omega_1 = \omega_2 = \omega. \tag{4.4}$$

In plain English, a sensible heating or cooling process is a constant-ω process. This, however, does not mean that the relative humidity, ϕ, stays the same. On the contrary, any increase in the dry-bulb temperature causes the air to become more thirsty. In other words, sensible heating boosts the air's ability to hold moisture. For that reason, the relative humidity decreases for this constant-absolute-humidity process, as the denominator, whereas, the maximum allowable humidity, which signifies its ability to absorb moisture, increases.

Without work or changes in potential and kinetic energy, following Eq. 4.1, the first law of thermodynamics can be expressed as

$$m_{a,1}'h_1 + Q' = m_{a,2}'h_2. \tag{4.5}$$

The sensible heating rate is thus,

$$Q' = m_a' (h_2 - h_1). \tag{4.6}$$

Since ω is constant, this can be written as

$$Q' = m_a' (c_{Pa} + c_{Pwv} \, \omega) (T_2 - T_1). \tag{4.7}$$

In practice, $(c_{Pa} + c_{Pwv} \, \omega)$ can be approximated to be equal to 1 kJ/kg·K.

4.2.2 Adiabatic Mixing of Two Streams of Moist Air

Figure 4.5 is a schematic representing a typical junction in the air handling system, where the return air at Location 1 merges with the required fresh air at Location 2, combining to become the supply air at Location 3. The conservation of the mass of dry air at the junction can be expressed mathematically as

$$m_{a,3}' = m_{a,1}' + m_{a,2}'. \tag{4.8}$$

The amount of moisture in the supply air is made up of that in the return air plus that in the fresh air, i.e.,

$$m_{a,3}' \omega_3 = m_{a,1}' \omega_1 + m_{a,2}' \omega_2. \tag{4.9}$$

The enthalpy (thermal energy) in the supply air (Location 3) is the sum of enthalpy in the return stream (Location 1) and fresh incoming air (Location 2),

$$m_{a,3}' h_3 = m_{a,1}' h_1 + m_{a,2}' h_2. \tag{4.10}$$

From the above conservation equations, it can be shown that (as an end-of-the-chapter problem),

$$m_{a,1}'/m_{a,2}' = (h_2 - h_3)/(h_3 - h_1) = (\omega_2 - \omega_3)/(\omega_3 - \omega_1), \tag{4.11}$$

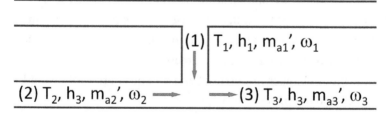

Figure 4.5. Fresh air plus return air to make the supply air (created by D. Ting). Adiabatic mixing of two streams of air at different conditions.

and hence,

$$(h_3 - h_2)/(\omega_3 - \omega_2) = (h_3 - h_1)/(\omega_3 - \omega_1). \qquad (4.12)$$

That is, the process from 1 to 3 has the same $\Delta h/\Delta\omega$ as that from 2 to 3, see Fig. 4.6. In plain English, the adiabatic mixing process is a constant-$\Delta h/\Delta\omega$ process. It is clear from Fig. 4.6 that

$$m_{a,1}'/m_{a,2}' = \text{Line } 2\text{-}3/\text{Line } 1\text{-}3. \qquad (4.13)$$

In other words, increasing the mass flow rate of Stream 1, $m_{a,1}'$, will bring State 3 closer to State 1, by extending Lines 2–3. Conversely, increasing the mass flow rate of Stream 2, $m_{a,2}'$, will move State 3 closer to State 2, i.e., Lines 1–3 increases.

The conservation of energy for the adiabatic mixing process can also be expressed as

$$m_{a,3}'(c_{Pa} + \omega_3\, c_{Pw})\, T_3 = m_{a,1}'(c_{Pa} + \omega_1\, c_{Pwv})\, T_1 + m_{a,2}'(c_{Pa} + \omega_2\, c_{Pwv})\, T_2. \qquad (4.14)$$

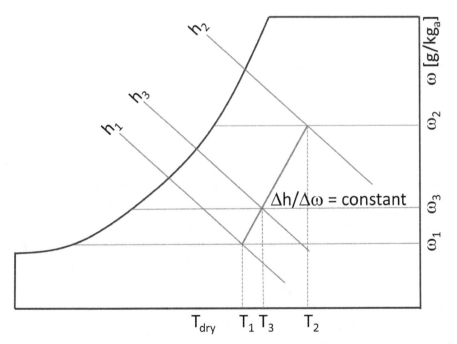

Figure 4.6. Adiabatic mixing process on a psychrometric chart (created by D. Ting). Note that the outgoing air stream after mixing is along a straight line connecting the states of the two incoming streams.

The terms in brackets are the specific heat capacities of the moist air in Stream 3, Stream 1, and Stream 2, respectively. Over a narrow range of conditions (temperature), these specific heat capacities may be approximated to be equal, i.e.,

$$c_{P,3} \approx c_{P,2} \approx c_{P,3}. \qquad (4.15)$$

Under these approximations, we have

$$T_3 \approx \frac{\dot{m}_{a,1}}{\dot{m}_{a,3}} T_1 + \frac{\dot{m}_{a,2}}{\dot{m}_{a,3}} T_2. \qquad (4.16)$$

This grants further evidence to the discussion on Eq. 4.13 concerning a larger mass leads to a larger influence. An increase in $m_{a,1}'$ will bring T_3 closer to T_1, and an increase in $m_{a,2}'$ will bring T_3 closer to T_2. Such is also the case in parenting. Namely, the parent who spends more time with the child, larger m_a', will draw the child closer to her or him.

As we are talking about engineering approximations, it is timely to convey the density or specific volume of standard air. Standard air corresponds to saturated air at 15°C (60°F), or dry air at 20°C (69°F). The density of standard air, $\rho = 1.204$ kg/m³ (0.075 lbm/ft³), or its inverse, i.e., the specific volume is $v_{std} = 0.830$ m³/kg (13.33 ft³/lbm). One advantage of using a psychrometric chart (and psychrometric software) is that these approximations need not be made, improving the accuracy of the calculations.

4.2.3 Heating with Humidification

Heating with humidification is imperative in engineering human thermal comfort during the winter months. For most effective humidification, one would heat up the air before applying humidification. This is the case in typical heating with humidification such as that realized via a residential furnace installed with a humidifier. In real applications such as this, however, only a portion of the heated air is passed through the humidifier. This is to prevent the thirsty heated air from picking up too much moisture. Analogously, when going for a long trip, parents do not inundate their children with water just because they are thirsty, as this will likely cause them to wet their pants somewhere along the trip, especially when passing through cold sissing brooks or musical rain. Thus, also we can imagine the air molecules; they become parched passing through the furnace. Fully quenching the temporal aridity will likely lead to precipitation somewhere along the long, cooler air distribution path, promoting mold formation and other damage.

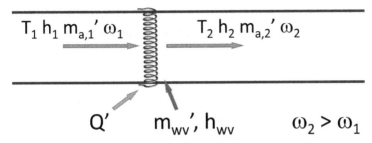

Figure 4.7. Heating with humidification (created by D. Ting).

Consider the heating with humidification process as portrayed in Fig. 4.7. The conservation of energy with $\Delta PE = \Delta KE = Work = 0$ gives

$$Q' + m_{a,1}{}'h_1 + m_{wv}{}'h_{wv} = m_{a,2}{}'h_2. \qquad (4.17)$$

Note that Q' is the sensible heating rate, whereas the energy input associated with the humidification is $m_{wv}{}'h_{wv}$. The conservation of dry air is as described by Eq. 4.2, i.e., the incoming dry air mass flow rate equals the outgoing one.

The conservation of moisture (water vapor) can be expressed as

$$m_{a,1}{}'\omega_1 + m_{wv}{}' = m_{a,2}{}'\omega_2. \qquad (4.18)$$

That is, the mass flow rate of the moisture associated with the incoming air plus that from humidification is equal to the outgoing water vapor mass flow rate. This can be rewritten as

$$m_{wv}{}' = m_a{}'(\omega_2 - \omega_1), \qquad (4.19)$$

which says that the mass flow rate of water vapor added via humidification is equal to the increase in the humidity ratio of the air stream multiplied by the mass flow rate of dry air.

Substituting Eq. 4.19 into the energy equation, Eq. 4.17, gives

$$Q' = m_a{}'[h_2 - h_1 - h_{wv}(\omega_2 - \omega_1)]. \qquad (4.20)$$

When there is humidification only, $Q' = 0$, we have

$$h_2 - h_1 = h_{wv}(\omega_2 - \omega_1). \qquad (4.21)$$

This can be recast as

$$h_{wv} = (h_2 - h_1)/(\omega_2 - \omega_1), \qquad (4.22)$$

which describes the increase (change) in enthalpy with respect to the increase in humidity ratio. The many different scenarios can be classified into the following types, see Fig. 4.8.

- Constant-T_{dry} humidification. For this kind of humidification, the water vapor added has a specific enthalpy, h_{wv}, that is equal to that of the saturated steam, h_g, at T_{dry}. Since T_{dry} remains fixed, there is no sensible heating, i.e., this is a pure humidification process.
- Proceeding to the right of the constant-T_{dry} humidification course signifies some sensible heating associated with the humidification. Namely, if the specific enthalpy of humidifying water vapor, h_{wv}, is greater than that of saturated steam at T_{dry}, h_g, the air will be sensibly heated as it is being humidified.
- If h_{wv} is less than h_g, the air will be cooled during the process of humidification. One special case of this is constant-T_{wet} humidification. Water at T_{wet} is used for humidifying the air. In this manner, the humidification process proceeds along the constant T_{wet} line. As this constant-T_{wet} course is accompanied with a decrease in T_{dry}, it bespeaks sensible cooling. A "cool mist" room humidifier operates in this manner.

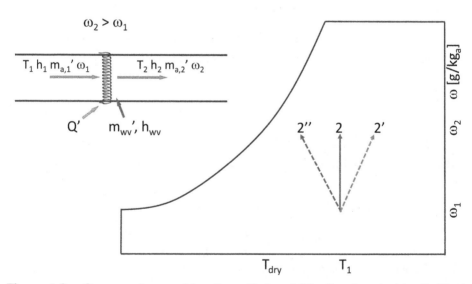

Figure 4.8. Common types of heating with humidification (created by D. Ting). Note that cold water with lower enthalpy than the air can in fact reduce the dry-bulb temperature, i.e., resulting in cooling instead of heating.

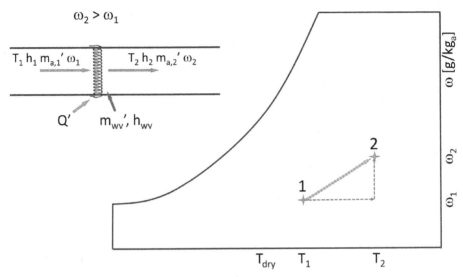

Figure 4.9. Treating heating with humidification as a two-part process, sensible heating followed by constant-T_{dry} humidification (created by D. Ting).

To ease calculation, we can divide the heating with dehumidification process into two processes, i.e., sensible heating and pure (constant-dry-bulb-temperature) humidification. This two-part process is illustrated in Fig. 4.9. The incoming air is first heated, and subsequently, humidified at a constant T_{dry}.

4.2.4 Cooling with Dehumidification

When moist air is cooled to or below its dew point, condensation of moisture will occur. A common example of this is the "leakage" underneath a car when running its air conditioning on a humid summer day. Consider the steady-state process illustrated in Fig. 4.10. The conservation of energy can be expressed mathematically as

$$Q' + m_{a,1}' h_1 = m_{a,2}' h_2 + m_f' h_{f,2}. \qquad (4.23)$$

Here, following the standard thermodynamic notation, subscript "f" signifies the liquid phase (liquid water), whereas subscript "g" has been used to denote the gaseous phase (water vapor or steam).

The conservation of dry air results in the same equation as Eq. 4.2. The conservation of moisture (water vapor) says that the mass flow rate of water vapor in the incoming moist air is equal to that in the outgoing moist air plus the rate of condensation, i.e.,

$$m_{a,1}{}' \omega_1 = m_{a,2}{}' \omega_2 + m_f{}'. \tag{4.24}$$

Ideally, the outgoing air is fully saturated whenever there is condensation. In plain English, the (bulk) air temperature reaches the dew point in the ideal cooling and dehumidification process. The moisture conservation equation can be rearranged to give the condensation rate, i.e.,

$$m_f{}' = m_a{}' (\omega_1 - \omega_2). \tag{4.25}$$

Substituting this into the energy equation and reorganizing yield

$$Q' = m_a{}' \left[h_2 - h_1 + h_{f,2} (\omega_1 - \omega_2) \right]. \tag{4.26}$$

This equation states that the heat into the system (air) is equal to the gain in the outgoing air enthalpy from that of the incoming air enthalpy, plus the enthalpy of the outgoing condensate. For the cooling and dehumidification process, the direction of the heat movement is negative, i.e., out of the system. Accordingly, the rate of heat removal is the decrease in the (moist) air enthalpy minus the enthalpy of the condensate.

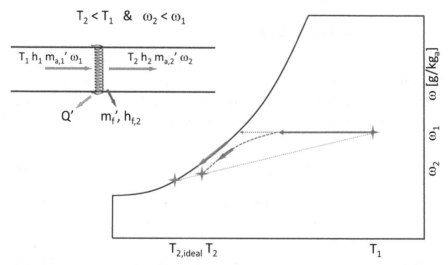

Figure 4.10. Cooling with dehumidification (created by D. Ting). Ideally, the air is sensibly cooled until it reaches saturation, and further cooling condenses moisture out of the cool air.

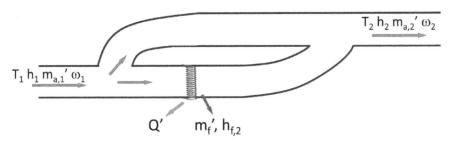

Figure 4.11. Bypassing cooling or heating coil (created by D. Ting). In reality, not all the air passing through a cooling or heating coil comes in contact with the coil so that its temperature approaches that of the coil surface.

Bypass Factor Let us expound on the earlier comment regarding inhomogeneity in practical cooling processes, i.e., where, practically, sensible-only cooling has to stop somewhat ahead of the dew point, below which some condensation materializes. Imagine Ms. Frizzle takes us in the Magic School Bus. After shrinking us to the fluid-parcel[2] size, we park on the cooling coil and follow the incoming hot air parcels. It is evident that the proximity of each passing fluid parcel to the cold cooling coil surface varies. Some fluid parcels literally touch the cold surface, whereas others pass by at varying distances. This continuous variation can be simplified into a two-way passage as portrayed in Fig. 4.11. The portion of air bypassing the coil is not cooled, whereas the remaining portion is completely cooled. The proportion of these two portions gives the equivalent average temperature of the air after passing through the coil section.

At this point, we can invoke the fundamental laws to the bypass case depicted in Fig. 4.11. The conservation of dry air gives

$$m_{a,1}' = m_{a,2}' = m_a'. \tag{4.27}$$

The bypass factor can be defined as

$$b = m_{a,b}'/m_a', \tag{4.28}$$

where $m_{a,b}'$ is the mass flow rate of the dry air bypassing the cooling or heating coil.

---

[2]A fluid parcel is a very small mass, or volume for incompressible flow, of fluid moving with the flow. It is large compared to the molecules and their mean free path, but small with respect to the length scales of the flow. It is conveniently used for describing the average velocity and other properties at the elemental level.

For the adiabatic mixing of the two streams, conservation of energy can be expressed as

$$m_{a,b}{'} c_{P,1} T_1 + m_{a,d}{'} c_{P,d} T_d = m_{a,2}{'} c_{P,2} T_2. \tag{4.29}$$

But $c_{P,1} \approx c_{P,d} \approx c_{P,2}$, and thus, the energy conservation expression may be approximated as

$$T_2 \approx (m_{a,b}{'}/m_a{'}) T_1 + (m_{a,d}{'}/m_a{'}) T_d, \tag{4.30}$$

or,

$$m_{a,b}{'}/m_a{'} \approx (T_2 - T_d)/(T_1 - T_d). \tag{4.31}$$

Substituting this into the bypass factor expression, we get

$$b = m_{a,b}{'}/m_a{'} \approx (T_2 - T_d)/(T_1 - T_d). \tag{4.32}$$

This repeats what we have discussed earlier. In this bypass cooling framework, the larger the amount of air bypassing the cooling coil, the closer to the outgoing air will be to the initial temperature. Therefore, measures such as cooling coils with large heat transfer surfaces populating the air passage section can be implemented to reduce the bypass factor, boosting the efficacy of the cooling.

It is also worth iterating the relative proportion of sensible and latent heating, noting that cooling is simply heating with the direction of thermal energy flow reversed. The total heat transfer rate in this context consists of the sensible portion and the latent part, i.e.,

$$Q_{tot}{'} = Q_S{'} + Q_L{'}. \tag{4.33}$$

The sensible heat transfer rate,

$$Q_S{'} = m_a{'} c_P (T_1 - T_2). \tag{4.34}$$

The **Sensible Heat Ratio** (SHR) that signifies the proportion of sensible heating to the sum of sensible and latent heating is

$$SHR = Q_S{'}/Q_{tot}{'}. \tag{4.35}$$

As mentioned earlier, a protractor is provided on the top left corner of a standard pyschrometric chart to ease the deduction of heating and cooling processes where the SHR is known or is required.

EXAMPLE 4.1. COOLING OF MOIST AIR

Given: Air at atmospheric pressure is cooled from $T_{dry} = 27°C$ and $T_{wet} = 24°C$ to saturation at $13°C$.

Find: (a) amount of water removed, (b) latent heat removed, (c) the SHR (sensible heat ratio)

Solution:

This problem can be easily solved with the help of a psychrometric chart. Initially, $T_{dry} = 27°C$ and $T_{wet} = 24°C$, from the psychrometric chart,

$$\omega_1 = 0.0178, h_1 = 72 \text{ kJ/kg}_a.$$

To saturation at $13°C$, on the psychrometric chart,

$$\omega_2 = 0.0095, h_2 = 38 \text{ kJ/kg}_a.$$

Thus,

$$\omega_1 - \omega_2 = 0.0083 \text{ kg}_w/\text{kg}_a.$$

State 1' on the psychrometric chart is from State 2 moving horizontally right until it reaches $T_{dry} = 27°C$, i.e., a constant humidity ratio process with $\omega = \omega_2 = 0.0095 \text{ kg}_w/\text{kg}_a$. Therefore,

$$h_{1'} = 51 \text{ kJ/kg}_a.$$

Latent heat removed $= h_1 - h_{1'} = 21 \text{ kJ/kg}_a$.
Sensible heat removed $= h_{1'} - h_2 = 13 \text{ kJ/kg}_a$.
In short, the sensible heat ratio, SHR $= 13/(21 + 13) = 0.38$.

4.2.5 Evaporative Cooling

Continuing along the topic of cooling, the most common and taken-for-granted evaporative cooling is doubtlessly sweating. In other words, when exposing more of our precious skin to allow largely sensible cooling is not enough to get rid of metabolized thermal energy, our body resorts to the substantially more-effective cooling, sweating. Our ancestors have long capitalized this potent cooling, effectively and simply engineering it into various technologies. For example, the

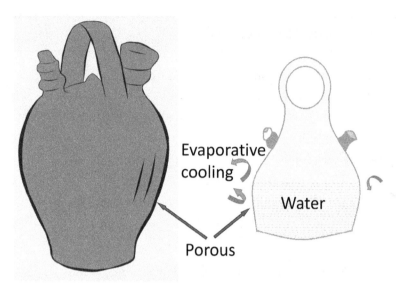

Figure 4.12. Evaporative cooling in practice (created by S. Akhand & S.K. Mohanakrishnan, edited by D. Ting). A small amount of water seeps through the porous container. Part of the latent heat of evaporation comes from the container-water system. As the system loses energy, it cools down.

Spanish have taken advantage of botijo, see Fig. 4.12, to evaporatively cool water under the hot sun. Similarly, Texas cowboys have utilized porous containers made of clay to cool drinking water via forceful evaporation in the desert-like environment. It is the dryness, along with the draft, that truly enhances the efficacy of evaporative cooling. Even in relatively humid South-Western Ontario, Canada, many greenhouses put into action evaporative cooling during the summer months, where water is drained down a corrugated grid in front of a fan. In this particular case, the cooling efficiency is presumably of secondary concern, for humidification is generally good for the greens inside the greenhouse.

Evaporative cooling is essentially an incomplete adiabatic saturation process, as illustrated in Fig. 4.13. The **Saturation Effectiveness** [Kreider & Rabl, 1994], which signifies the performance of evaporative cooling, can be expressed as

$$e_{sat} = (T_1 - T_2) / (T_1 - T_{sat}).\qquad(4.36)$$

Alternatively, the effectiveness can be described by the **Evaporative Cooling Efficiency**,

$$\eta_{EC} = (\omega_2 - \omega_1) / (\omega_{sat} - \omega_1).\qquad(4.37)$$

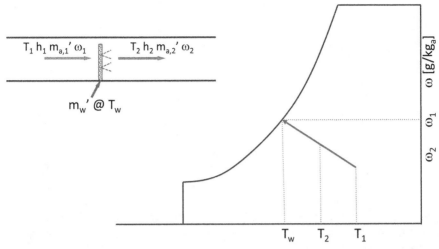

Figure 4.13. Saturation effectiveness of evaporative cooling (created by D. Ting).

For $c_P \approx$ constant, the thermal energy change $\approx m \, c_P \, \Delta T$, where ΔT is the drop in the dry-bulb temperature. With this approximation, we have

$$\eta_{EC} = (T_1 - T_2) / (T_1 - T_{sat}).\tag{4.38}$$

4.3 CONDITION LINE FOR A SPACE

With the essential psychrometric processes under our belt, we can look at the condition line for a space. A condition line is a line on a psychrometric chart that illustrates the required cooling with or without dehumidification, or, heating with or without humidification. For a typical space, as shown in Fig. 4.14, the supply air picks up or rejects sensible and latent heat from the space and exits the space as return air. Simply put, the condition line indicates the total amounts of sensible and latent heat a space gains (or loses) on a psychrometric chart. These need to be removed (or replenished) as realized by a suitably sized HVAC system.

Typical energy gains and/or losses include the following:

1. Those transferred across the building envelope, i.e., via convection-conduction-convection heat transfer
2. Internal gains
3. Solar gain
4. Infiltration/exfiltration

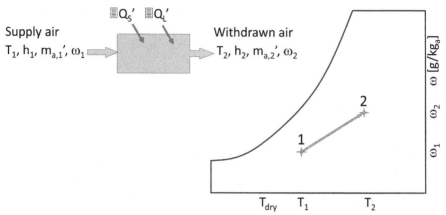

Figure 4.14. From supply to return air (created by D. Ting). This is the case during the summertime, where cooled air entering the occupied space picks up both sensible and latent heat.

Note that the infiltration needs to balance the exfiltration, $m_{a,1}' = m_{a,2}' = m'$, so that the space is neither over-pressurized nor under-pressurized. Over pressurization will force the nicely conditioned air through the various cracks (openings) of the building envelope. On the other hand, under pressurization may result in undesirable infiltration of unconditioned outdoor air.

For the steady-flow situation depicted in Fig. 4.14, the conservation of energy with $\Delta PE = \Delta KE = Work = 0$ gives

$$m_{a,1}' h_1 + \sum Q_S' + \sum Q_L' = m_{a,2}' h_2. \tag{4.39}$$

The incoming enthalpy of the supply air plus all the gains in terms of both sensible and latent heat is equal to the outgoing return air enthalpy. The conservation of dry air mass leads to

$$m_{a,1}' = m_{a,2}' = m_a'. \tag{4.40}$$

With this, the energy equation can be simplified into

$$\sum Q_S' + \sum Q_L' = m_a' (h_2 - h_1). \tag{4.41}$$

The total sensible heat gains plus the sum of all latent heat gains is equal to the change in the enthalpy of the air from the supply to the return stream. The moisture must also be conserved, i.e.,

$$m_{a,1}' \omega_1 + \sum m_{wv}' = m_{a,2}' \omega_2. \tag{4.42}$$

This can be recast into

$$\sum m_{wv}' = m_a'(\omega_2 - \omega_1). \qquad (4.43)$$

That is, the total moisture gained in the space is equal to the increase in the humidity ratio from supply to return, multiplied by the mass flow rate of the circulating air. From these, we can write the enthalpy–water vapor ratio as

$$(h_2 - h_1)/(\omega_2 - \omega_1) = \left(\sum Q_S' + \sum Q_L'\right)/\sum m_{wv}'. \qquad (4.44)$$

We will illustrate the utilization of the space condition line next.

4.3.1 Single-Zone Summer Cooling

A zone is an area of a building that requires a specific comfort condition. As such, it has its own thermostatic and at times humidity control. A simple residential building is probably the most common single zone in practice, where only one thermostat is employed to control the indoor temperature of the entire building. While small, portable humidifiers may be utilized for special needs of particular rooms or areas, most houses have effectively only one humidifier installed onto the main ductwork of the furnace to mitigate discomfort associated with dryness imposed by the heating of cold air during the winter months. Typical operation of the single-zone residential air conditioning, as illustrated in Fig. 4.15, does not involve humidity control.

Let us examine the single-zone cooling and dehumidification process depicted in Fig. 4.15. Line 1–2 is the space-condition line, where the entering cool air picks up the net sensible and latent heat gains and returns to the cooling system. A small portion of the air exits the building as exhaust air, and this portion can become quite significant when exhaust fans are turned on high. Some fresh air, via infiltration and/or an air-to-air heat exchanger, enters at Point 4 to balance the exhaust and/or exfiltration. It is clear that Line 4–2–5, which denotes the mixing of fresh air that enters at Point 4 with return air from Point 2 at Point 5, is an adiabatic mixing line. This mixed air stream is cooled by the cooling coil, and normally some condensation is formed. Accordingly, Line 5–6 is a cooling and dehumidification line. With a negligible pressure rise through the air distribution fan, the condition of air at Point 1 is equal to that at Point 6. Note that the air at these points is cold and damp. In practice, some warming, i.e., sensible heating, along the air duct pushes Point 1 slightly to the right on the psychrometric chart. The air temperature rises to the comfortable range with the indoor (space) heat gains.

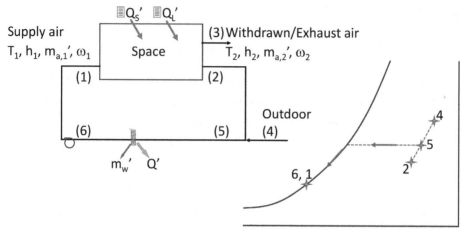

Figure 4.15. Single-zone summer cooling with dehumidification (created by D. Ting).

Naturally, a certain amount of bypassing occurs through the cooling coils. This is another reason why condensation does not generally take place farther downstream of the cooling coil in the summer time. In higher-end houses and many commercial buildings, more deliberate humidity management is made possible by implementing a physical bypass. In such a case, a larger and controllable portion of the returning air is directed through a bypass duct, which is "psychrometrically" parallel to the duct encompassing the cooling coil. Also resorted to is administered reheating after the cooling coil. This reheating is somewhat wasteful as far as energy usage is concerned. This is because some amount of additional heat is needed, and this adds to the cooling requirement. Nonetheless, this reheating approach is decidedly more effective in controlling both the temperature and the humidity.

EXAMPLE 4.2. SINGLE-ZONE SUMMER COOLING

Given: An occupied space gains sensible heat at 110 kW and moisture at 0.04 kg/s as depicted in Fig. 4.15. Air leaves the cooling coil at 12°C and 90% relative humidity (if needed, you may assume $h = 2560$ kJ/kg$_w$, $c_P = 1$ kJ/kg·K).

Find: (a) T_{dry} and T_{wet} of the supply air, (b) rate of heat input via the reheat coil, (c) locate all points on the psychrometric chart provided.

Solution:

This problem can be easily solved with the help of a psychrometric chart.

Cooling season, summer indoor design/desirable conditions, $\approx 24°C$ and $\approx 50\%$.

$$Q_{latent}{}' = m_w{}'h = 0.04\,(2560) = 102.4\,kW$$

$$SHR = 110/(110 + 102.4) = 0.52$$

State 6: 12°C and 90% relative humidity

State 1 is at the intersection of SHR = 0.52 through State 2 and constant ω from State 6

From psychrometric chart,

$$T_1 = 20.5°C \,\&\, T_{wet,1} = 14.5°C$$

$$m_{a,1}{}' = Q_{latent}{}'/[h\,(\omega_2 - \omega_1)] = 102.4\,kW/[2560\,(0.0093 - 0.008)] = 30.8\,kg/s$$

$$Q_{6-7}{}' = m_{a,1}{}'\,c_P\,(T_1 - T_6) = 30.8\,(1)\,(20.5 - 12) = 262\,kW$$

4.3.2 Single-Zone Winter Heating

For the winter, we typically have both sensible and latent heat losses, where the loss of moisture is largely due to the exchange of indoor air with dry outdoor air. Therefore, we usually have to supply air at a higher temperature and humidity (ratio) than the air in the space. This is schematically shown in Fig. 4.16. Line 1–2 is the space-condition line, showing that both sensible heat and moisture are lost in the space. Point 4 corresponds to the cold and dry (low absolute humidity) outdoor air. For the occupants to stay alive, some minimal amount of this fresh air must enter the building. As such, Line 4–2–5 is the now-familiar adiabatic mixing line, adiabatically mixing the return or recirculating air with some fresh outdoor air. Anticipating heat loss farther downstream, especially from the relatively large space, which is the entire building in its simplest form, the mixed recirculating air is sensibly heated above the comfortable temperature. This is depicted by Line 5–6, which is a constant-ω line. For the case in which room temperature water is used for humidification, Line 6–7 is a constant-T_{wet} line, which is an ideal humidification process.

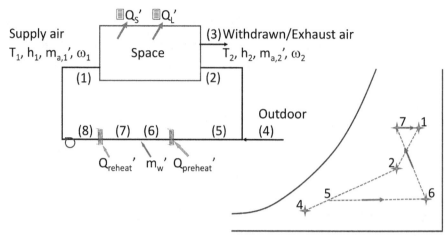

Figure 4.16. Single-zone winter heating and humidification (created by D. Ting). In this particular case, preheating is employed to enhance humidification, and reheating is utilized to control the humidity of the heated air.

It is clear that Point 7 is in the proximity of saturation. To prevent condensation along the relatively cooler air duct downstream, it must be heated to move it away from the saturation line. This is generally done via reheating the humidified air sensibly, i.e., the constant-ω Line 7–8. With negligible changes across the fan, Point 1 is equal to Point 8 for all practical purposes.

It is worth emphasizing that the preheat coil helps in preventing water in the air washer (humidifier, after the air is filtered by an air filter) from the possibility of freezing. Primarily, it regulates the amount of water evaporated in the air washer. The reheat coil, on the other hand, controls the dry-bulb temperature of the supply air. Indirectly, this regulates the supply air humidity, and thus, prevents mold formation and other high-humidity-related damage.

4.3.3 Multi-Zone Cooling and Heating

It is often necessary to divide a building into different zones due to (a) spatial variations of heating and cooling loads and (b) different usage for different areas and hence, different conditions for thermal comfort. For example, a building that houses a swimming pool along with offices would definitely require at least two very different zones. The swimming pool zone would be warm and humid, whereas the office area should be much cooler and drier. Even for an office-only building, different zones are generally needed. An obvious case is an office building with beautiful and soul-soothing facades. The offices facing the sun at any one time would typically have a substantially larger heat gain. As a result, the supply air for these sunlit rooms has to be cooled to a lower temperature and/or be supplied at a higher flow rate during the warm season. In the winter time, a lower volumetric supply of cooler or less-heated air is needed. On the other side, the space shaded from the sun may require heating, e.g., from midnight until morning during the transitional seasons, i.e., spring and fall.

PROBLEMS

Problem 4.1

Concerning the common psychrometric chart.
The horizontal axis is _____
The vertical axis is _____
The two parameters whose contour lines have very similar negative slopes are
_____ and _____

Problem 4.2

Show that, using the conservation laws, the adiabatic mixing of Stream 1 and Stream 2 into Stream 3 can be described as

$$m_{a,1}'/m_{a,2}' = (h_2 - h_3)/(h_3 - h_1) = (\omega_2 - \omega_3)/(\omega_3 - \omega_1).$$

Problem 4.3

Outdoor air at $-5°C$ and 60% relative humidity is heated and humidified by saturated steam at 100°C. The airflow rate is 8000 L/s, and heat is added to the air at the rate of 300 kW while it absorbs 0.05 kg/s of steam. What are the dry-bulb and wet-bulb temperatures of the air leaving the heating and humidifying system?

Problem 4.4

Outdoor air with $T_{dry} = 35°C$ and $T_{wet} = 20°C$ enters an air conditioner's cooling coil at 1000 L/s. The air is cooled to 12°C and 90% relative humidity, with liquid water leaving the cooling coil at 10°C. What is the coil heat transfer rate?

Problem 4.5

Air enters an evaporative cooler at 40°C and 20% relative humidity. It is humidified to 80% T_{dry} depression (saturation effectiveness) at 2000 L/s. What is the water evaporation rate?

Problem 4.6

An occupied space gains sensible heat at 110 kW and moisture at 0.04 kg/s. Air leaves the cooling coil at 12°C and 90% relative humidity. (If needed, you may assume $h = 2560$ kJ/kg$_w$, $c_p = 1$ kJ/kg·K). Find T_{dry} and T_{wet} of the supply air, and the rate of heat input via the reheat coil.

REFERENCES

J.F. Kreider, A. Rabl, *Heating and Cooling of Buildings: Design for Efficiency*, Chapter 4, McGraw-Hill, New York, 1994.

R. Mollier, A new diagram for water vapor-air mixtures, Mechanical Engineering, 45: 703–705, December 1923.

CHAPTER 5

Building Heat Transmission

"Heat, like gravity, penetrates every substance of the universe, its rays occupy all parts of space."

–Jean-Baptiste-Joseph Fourier

CHAPTER OBJECTIVES

- Review the three heat transfer mechanisms.
- Recapitulate the concepts of thermal resistance, thermal diffusivity, and heat capacity.
- Recognize emissivity, absorptivity, transmissivity, and reflectivity in radiation heat transfer.
- Learn to practically handle heat transfer across a cavity, filled with gas (air).
- Appreciate the difference in parallel-path and isothermal-plane approximations in analyzing heat transfer across multilayered building structures.
- Understand how to estimate heat transfer through windows.
- Become familiar with heat loss through basement walls and floors.
- Comprehend the transport of moisture through a building envelope.

Nomenclature

A   Area; A_{avg} is the average area, A_{cg} is the area of the center of glass window, A_{eg} is the area of the edge of the glass window, A_f is the area of the frame of the glass window, A_s is the surface area, A_{stud} is the area of the stud

c   Heat capacity; c_{iron} is heat capacity of iron, c_P is heat capacity at constant pressure

(Continued)

E   Energy; E_b is the energy emitted by a blackbody, E_{in} is the energy entering a system, E_{out} is the energy leaving a system, ΔE_{sys} is the energy change of a system

F   Shape factor

h   Heat transfer coefficient; h_{conv} is the convection heat transfer coefficient, h_{rad} is the radiation heat transfer coefficient

HVAC   Heating, ventilation, and air conditioning

I   Current

k   Thermal conductivity; k_{fluid} is the thermal conductivity of the fluid, k_{gas} is the thermal conductivity of the gas

KE   Kinetic energy; ΔKE is the change in kinetic energy

L   Length or thickness

M   Moisture transport conductance, $M = \mu/L$

m   Mass; $m_w{}'$ is the mass flow rate of water vapor, m_{water} is the mass of water

P   Pressure; P_w is the partial pressure of water vapor, $P_{w,cold}$ is the partial pressure of water vapor on the cold side, $P_{w,sat}$ is the partial pressure of water vapor at saturation, $P_{w,warm}$ is the partial pressure of water vapor on the warm side, $P_{w,x}$ is the partial pressure of water vapor at location x

PE   Potential energy; ΔPE is the change in potential energy

Q   Heat or thermal energy; Q' is the heat transfer rate, $Q_{absorbed}{}'$ is the rate of radiation being absorbed, $Q_b{}'$ is the heat transfer rate through the board, $Q_{cond}{}'$ is the conduction heat transfer rate, $Q_{conv}{}'$ is the convection heat transfer rate, $Q_{emit\,max}{}'$ is the maximum radiation power emitted by a blackbody, $Q_{incident}{}'$ is the rate of radiation incident on a surface, $Q_{overall}{}'$ is the overall heat transfer rate, $Q_{rad}{}'$ is the radiation heat transfer rate, $Q_{ref}{}'$ is the reflected radiation heat transfer rate, $Q_{stud}{}'$ is the heat transfer rate through the stud

q"   Heat flux

R   Resistance; R_{cond} is the thermal conduction resistance, R_{elec} is the electric conduction resistance, R_i is the inner or indoor thermal resistance, R_o is the outer or outdoor thermal resistance, R_{rad} is the radiation thermal resistance, R_{tot} is the total thermal resistance, R_{w-x} is the thermal resistance from the warm side to location x

r   Radius; r_i is the inner radius, r_o is the outer radius

(*Continued*)

R$_{th}$ R-value, $R_{th} = \Delta x/k = AR$

SI   International System of Units

T   Temperature; T_A is the amplitude of the ground surface temperature or deviation from the deep soil temperature, T_{avg} is the average temperature, T_{cold} is the temperature of the cold side, T_{eqm} is the temperature at equilibrium, T_g is the ground temperature, T_H is the high temperature, T_i is the inner or indoor temperature, T_L is the low temperature, T_o is the outer or outdoor temperature, T_s is the temperature of a surface, T_{sur} is the temperature of the surroundings, T_{warm} is the temperature of the warm side, T_x is the temperature at location x, T_∞ is the ambient temperature, ΔT is temperature difference

U   U-value, or, internal energy; U_{avg} is the average U-value, U_{cg} is the U-value of the center of the glass window, U_{eg} is the U-value of the edge of the glass window, U_f is the U-value of the frame of the glass window, U_o is the overall U-value, ΔU is the change in internal energy, ΔU_{iron} is the change in internal energy of iron, ΔU_{water} is the change in internal energy of water

V   Voltage; ΔV is the voltage difference, V_H is the high voltage, V_L is the low voltage

x   Distance in the x-direction; Δx is the thickness

y   Distance in the y-direction

Z   Water vapor-flow resistance, $Z = 1/M$; Z_{tot} is the total vapor flow resistance, $Z_{tot\text{-}vr}$ is the sum of all the vapor flow resistance excluding that of the vapor retarder, Z_{vr} is the required vapor flow resistance, $Z_{w\text{-}x}$ is the vapor flow resistance from the warm side to location x, $Z_{x\text{-}c}$ is the total vapor resistance from location x to the cold side

Greek and Other Symbols

α Thermal diffusivity, $\alpha = k/\rho c_p$

α_{rad} Absorptivity

μ The permeability of water vapor

ρ Density; ρ_{water} is the density of water

ρ_{rad} Reflectivity

σ Stefan–Boltzmann constant, $\sigma = 5.669 \times 10^{-8}$ W/(m$^2 \cdot$ K^4)

τ_{rad} Transmissivity

(*Continued*)

    ε Emissivity; ε_{eff} is the effective emittance, ε_H is the emissivity of the high-temperature surface, ε_L is the emissivity of the low-temperature surface

    $\forall$ Volume; $\forall_{water}$ is the volume of water

5.1 THE THREE HEAT TRANSFER MECHANISMS

The three fundamental heat transfer modes are (a) conduction, (b) convection, and (c) radiation (see Fig. 5.1). More fundamentally, the physics behind these three mechanisms can be described as follows [Bergman et al., 2011; Çengel & Ghakar, 2015]:

1) Conduction. In conduction, heat, or more correctly, thermal energy, is transferred via the molecular-level kinetic energy from the more energetic molecules to the less energetic ones. As such, solids where the molecules are most closely packed, especially metals, are better conductors.

2) Convection. Convection takes place in the form of larger-scale motions of a fluid, either liquid or gas. The more intense the flow motions, the higher the heat convection rate.

3) Radiation. Radiation involves the transport of thermal energy by electromagnetic waves. The large-vacuum space permits the radiation from the sun to travel far and wide, keeping our good planet from forming into an ice ball.

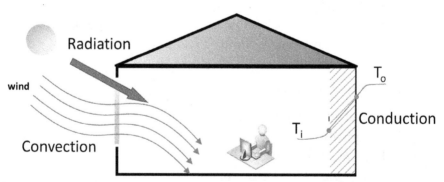

Figure 5.1.   Heating, ventilation, and air conditioning (HVAC) heat transfer involving the three constituent mechanisms: conduction, convection, and radiation (created by X. Wang, edited by D. Ting).

As this book is meant to be best utilized when complemented with the practical reference, the ASHRAE Handbook, readers are recommend to consult Chapter 4 of 2017 ASHRAE Handbook [ASHRAE, 2017] when they are practicing in the heating, ventilation, and air conditioning (HVAC) field. The handbook is particularly profitable for looking up the required values for complicated engineering components involved in the HVAC industry.

5.1.1 Steady, One-Dimensional Heat Conduction

Let us consider the one-dimensional case with constant thermal properties as depicted in Fig. 5.2, where heat is being conducted through a plane wall. Fourier's law of heat conduction states that the conduction heat transfer rate,

$$Q' = -kA \, dT/dx, \tag{5.1}$$

where k is the **thermal conductivity** in units of W/(m · K), A is the area, T is the temperature, and x is the distance. The negative sign implies that the heat is transferred from a high temperature to a low temperature region. This can be expressed more clearly as

$$Q' = kA \, (T_H - T_L)/\Delta x, \tag{5.2}$$

where the negative sign is taken care of by subtracting the lower temperature, T_L, from the higher temperature, T_H.

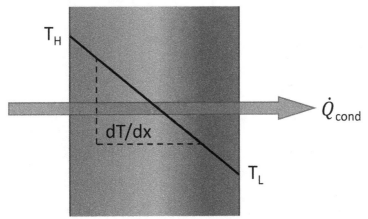

Figure 5.2.   Conduction heat transfer through a plane wall (created by D. Ting).

5.1.2 The Thermal Resistance Concept

Like an electric circuit of which the current flow rate is dictated by the electric resistance, the heat transfer rate depends on the heat transfer resistance involved (see Fig. 5.3). For conduction heat transfer through a layer of homogeneous material, Eq. 5.2 can be rewritten as

$$Q' = (T_H - T_L)/(\Delta x/kA) \tag{5.2a}$$

Introducing the **heat transfer resistance**,

$$R = \Delta x/(kA), \tag{5.3}$$

the one-dimensional heat conduction expression can be cast as

$$Q' = (T_H - T_L)/R. \tag{5.4}$$

Note that the heat transfer (thermal) resistance, R, is adopted, via analogy, from the electric resistance, which serves as the proportionality constant between voltage and current in Ohm's law. Also, like electric potential,

$$\Delta V = V_H - V_L, \tag{5.5}$$

and the corresponding thermal (temperature) potential,

$$\Delta T = T_H - T_L. \tag{5.6}$$

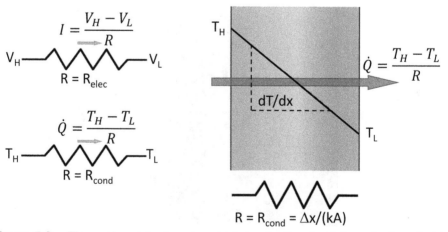

Figure 5.3. Thermal-resistant concept of conduction heat transfer (created by D. Ting).

Finally, the heat transfer rate is akin to the electric current, i.e.,

$$Q' = I. \tag{5.7}$$

The commonly used **R-value** is the unit thermal resistance,

$$R_{th} = \Delta x/k = AR. \tag{5.8}$$

Building materials such as insulation come with the appropriate R-value. The inverse of the R-value is the unit thermal conductance called the **U-value**, i.e.,

$$U = 1/R_{th} = k/\Delta x. \tag{5.9}$$

EXAMPLE 5.1. THERMAL CONDUCTIVITY AND THICK AND THIN WINTER COATS

Given: Professor Ting wears a bulky winter coat with a thickness of 3 cm to keep him warm. He sees Student Rich wearing a slick coat of one third the thickness of his coat. On a −22°C calm winter day, Student Rich lets Professor Ting try his thin coat. Professor Ting finds that the copacetic thin coat keeps him equally warm as his unsightly thick coat.

Find: What can you say about the thermal conductivity and thermal resistance of the two coats?

Solution:

The thermal conductivity, k, quantifies how well a material conducts heat. Assume that the outer surface of the coat is at −20°C (2°C warmer than the ambient due to the film thermal resistance), and the inner coat surface is at 20°C. Applying Eq. 5.2,

$$Q' = kA\,(T_H - T_L)/\Delta x = k_{3\,cm}\,A\,(20 + 20)\,°C/3\,cm$$
$$= k_{1\,cm}\,A\,(20 + 20)\,°C/1\,cm, \tag{5.1.1}$$

we see that the thermal conductivity of the thin coat, $k_{1\,cm}$, is one third that of the thick coat, $k_{3\,cm}$, and thus, the significantly higher price!

In terms of thermal resistance, $R = \Delta x/(kA)$, the three times thicker ($\Delta x = 3$ cm) coat has a three times larger k; therefore, $R_{3\,cm} = R_{1\,cm}$. In other words, the thin coat provides the same amount of thermal resistance as that of the thick coat.

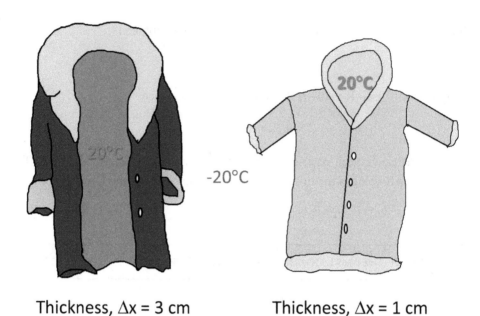

Thickness, Δx = 3 cm   Thickness, Δx = 1 cm

Figure 5.4.   Conduction through the same thermal resistance but different thickness winter coats (created by S.K. Mohanakrishnan).

Figure 5.4 illustrates the underlying heat transfer physics behind Example 5.1. Note that the outer and inner temperatures are the same for the thick and thin coats. Moreover, the rate of heat transmitted through heat conduction is the same for both coats. The fall out is that the thermal resistance of the thick coat equals to that of the thin one. In other words, the better coat can provide the same amount of insulation with a much thinner layer. This better coat does so by maintaining a significantly higher temperature gradient. We may imagine Mr. Freeze is on the outside trying to freeze the inside using his ice gun. The stronger temperature barrier requires a much thinner wall to keep Mr. Freeze and the freezing in check, maintaining a warm temperature inside in spite of significant outside "temperature pressure" or "temperature suction."

There was an engineering myth, where some researchers discovered that spider silk is more conductive than copper. This was presumably allured by the fame of The Amazing Spiderman comics, which uses the impressively strong spider silk to commute. The scientific paper was published, and then, it became a laughing-stock after arm-length researchers failed to duplicate the same result. Good engineers and good engineering researchers alike, must earn the trust of the public by being careful with the truth in all matters all the time. No wonder Albert Einstein asserted, "Whoever is careless with the truth in small matters cannot be trusted with important matters."

5.1.3 Thermal Diffusivity and Heat Capacity

Another parameter of importance is the thermal diffusivity, which signifies the heat capacity of a material. Specifically, the thermal diffusivity of a material,

$$\alpha = k/\rho c_p, \tag{5.10}$$

where ρ is the density of the material and c_p is the heat capacity. Together, the product, ρc_p, bespeaks how much thermal energy a unit volume of the material stores. Accordingly, α signifies the amount of thermal energy conducted through with respect to that stored within the material.

EXAMPLE 5.2. THERMAL DIFFUSIVITY AND THE PASSING VERSUS RETENTION OF THERMAL ENERGY

Given: Student Tech shows up with two identical budget winter coats. She replaces the fillings (between the inner and outer layers) of one of the coats with a lower thermal diffusivity material.

Find: What can you say about the performance of the coat with the lower thermal diffusivity material with respect to the other one?

Solution:

The thermal diffusivity, $\alpha = k/\rho c_p$, indicates how well a material lets the heat through with respect to retaining it. Consequently, the coat with the lower thermal diffusivity material performs better since it lets less heat through.

We may imagine the layer of the insulation as a row of miniature Engineering Human Thermal Comfort students. For Example 5.2, we have two rows with the same number (say 30) of students (same thickness), for both coats. Assume that Professor Ting brought two buckets filled with heated "hard cold cash," i.e., 100 Canadian toonies, which have been warmed up to 40°C in each bucket. Professor Ting's idea was to pass the buckets of warm toonies through the two rows of students to the needy people waiting in the cold. The row of students corresponding to the high thermal diffusivity coat can be envisioned to be less greedy, and/or more generous, than the one representing the low thermal diffusivity coat. If each student retains one toonie, there will be 70 toonies passed to the underprivileged group. Relatively, the low thermal diffusivity row keeps 90 toonies, i.e., 3 toonies per student, leaving only 10 toonies for the very needy ones.

(Continued)

Summarily, the material with a high thermal diffusivity grants easy passage to thermal energy, warm toonies in the aforementioned analogy. The low thermal diffusivity, $\alpha = k / \rho c_p$, has a relatively large denominator, ρc_p, i.e., the material is good at storing (or stealing) the thermal energy it receives. Consequently, very little thermal energy passes through it.

Other than density, ρ, the heat capacity (at constant pressure), c_p, (or simply c for incompressible materials), represents the amount of thermal energy needed to raise a unit mass of a material by one degree in temperature. A high heat capacity material can receive a lot of thermal energy without a significant rise in its temperature. Likewise, an individual with a high heat capacity can take on a lot of insult (e.g., slapping) before the person loses her/his cool.

EXAMPLE 5.3. HEAT CAPACITY: COOLING IRON WITH WATER

Given: A 50-kg iron block at 80°C is dropped into an insulated tank with 0.05 m³ of water at 25°C.

Find The equilibrium temperature, T_{eqm}, of the iron-in-water system.

Solution:

According to the conservation of energy (first law of thermodynamics),

$$E_{in} - E_{out} = \Delta E_{sys}. \tag{5.3.1}$$

This can be written as

$$Q - W = \Delta U + \Delta KE + \Delta PE. \tag{5.3.2}$$

For the closed system, we have

$$\Delta U = 0. \tag{5.3.3}$$

That is,

$$\Delta U_{iron} + \Delta U_{water} = 0. \tag{5.3.4}$$

For the incompressible solid and liquid, we can write

$$[m\,c\,(T_2 - T_1)]_{iron} + [m\,c\,(T_2 - T_1)]_{water} = 0. \tag{5.3.5}$$

For the water,

$$m_{water} = \rho_{water} V_{water} = 50\,kg.$$

From thermodynamics or material properties tables,

$$c_{iron} = 0.45\,kJ/kg \cdot {}^\circ C, \text{ and } c_{water} = 4.18\,kJ/kg \cdot {}^\circ C.$$

Note that water has a much higher heat capacity.
Substituting these values into Eq. 5.3.1, we get

$$T_{eqm} = T_2 = 30.3{}^\circ C.$$

In other words, though there are equal amounts of iron and water, the equilibrium temperature is much closer to that of the initial temperature of the water. This is because water has a much higher heat capacity, as noted above. With its higher heat capacity, it is less influenced by its surroundings, the hot iron in its midst, in this case.

It is thus clear that one should not consider iron when it comes to thermal energy storage. Readily-available water, on the other hand, is a good candidate for storing thermal energy. It is no wonder that some creative minds have resorted to water contained within transparent walls as a part of their building envelope. The large volume of water, which has a high heat capacity, furnishes a large thermal mass, flattening out daily variation in temperature. The resulting benefit is two-folds, simultaneous reduction in the midnight to early morning heating and minimization of afternoon cooling.

5.1.4 Steady Conduction in Cylindrical Coordinates

Cylindrical pipes and ducts are an integral part of heating, ventilation, and air conditioning (HVAC) systems. For a long cylindrical duct or pipe as shown in Fig. 5.5, the steady conduction rate can be expressed as

$$Q' = (T_i - T_o) / [\ln(r_o/r_i) / (2\pi kL)]. \tag{5.11}$$

Here, r_o is the outer radius, r_i, the inner radius, and L is the length of the pipe.

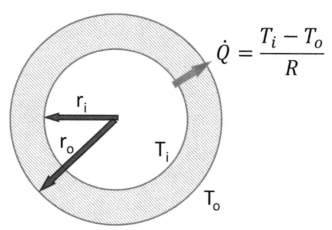

$$\dot{Q} = \frac{T_i - T_o}{R}$$

Figure 5.5. One-dimensional heat conduction through the wall of a long cylindrical pipe (created by D. Ting).

5.1.5 Heat Convection

When a moving fluid contacts a surface at a different temperature, the resulting heat transfer is called convection, see Fig. 5.6. In plain English, heat convection is the thermal energy transfer between a solid surface and the adjacent fluid that is in motion. As typical flow stops at a solid surface under the no-slip condition, the transfer of thermal energy at (next to) the surface occurs via conduction, in the absence of fluid motion. As a result, heat convection involves the combined effects of conduction and fluid motion.

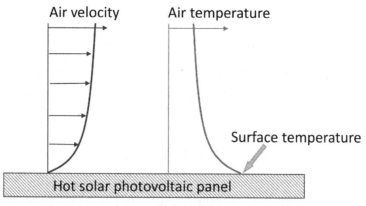

Figure 5.6.   Convection heat transfer (created by D. Ting).

Figure 5.7.   Natural versus forced heat convection (created by S. Akhand).

Convection heat transfer can be divided into two groups (Fig. 5.7).

1) Forced convection. When the fluid is forced to flow over a surface by an external means such as the ever-prevailing natural wind, a fan, or a pump.
2) Natural convection. When the fluid motion is caused by buoyancy force, the heated fluid expands and lightens, and thus, hot air rises. As the hot air rises, the cooler air rushes into the "vacant" region, and this results in natural convection-induced flow motion. Note that the buoyancy forces are only in effect under finite gravity, i.e., natural convection diminishes to zero as gravity approaches zero.

It is important to note that the no-slip condition at the wall typically makes up the largest thermal resistance when it comes to heat convection. According to Fig. 5.6, the rate of heat convection may be envisioned as two or more thermal resistors in series. Typically, one represents the convecting fluid resistance and the other the conduction via the non-moving fluid next to the solid surface. Without fluid motion, the resistor next to the solid surface has the largest resistance. Under steady-state conditions, the rate of heat passing through the series of resistors remains unchanged. For the heated solid surface case, the heat that passes through the second (convective) resistor is received from the first (conductive) one, i.e.,

$$Q_{conv}' = Q_{cond}' = -k_{fluid}\,(\partial T/\partial y)_{y=0}\,. \tag{5.12}$$

Recall that the negative sign manifests because heat flows in the negative temperature gradient direction, i.e., from high to low. Figure 5.6 shows that y is the distance in the vertical direction, where y = 0 is at the solid surface.

According to Newton's law of cooling,

$$Q_{conv}' = h_{conv}\,A_s\,(T_s - T_\infty)\,. \tag{5.13}$$

Here h_{conv} denotes the convective heat transfer coefficient, in SI units it is in $W/m^2 \cdot K$, A_s is the heated surface area at temperature T_s, and T_∞ is the temperature of the ambient fluid. If the surface is cooler than the surrounding fluid, then, the heat transfer direction is reversed. We can express the heat transfer rate as

$$Q_{conv}' = (T_s - T_\infty)/R, \tag{5.14}$$

where the **convection heat transfer resistance**,

$$R = 1/(h_{conv}\,A_s)\,. \tag{5.15}$$

The **thermal resistance value**, R_{th},

$$R_{th} = 1/h_{conv}, \tag{5.16}$$

and its reciprocal, the **U-value**,

$$U = h_{conv}\,. \tag{5.17}$$

5.1.6 Radiation Heat Transfer

In radiation, the transfer of thermal energy is via electromagnetic waves (photons) caused by variations in the electronic configurations of the atoms or molecules. As such, it is most effectively transmitted in the absence of an intervening medium. Being transmitted by electromagnetic waves, radiation has a spectrum of wavelengths and/or frequencies. Therefore, radiation heat transfer occurs by electromagnetic waves with a strong spectral dependence. The microscale details such as color of surfaces have a first-order effect on the rate of heat transfer. Also, radiation heat transfer is a highly non-linear function of temperature.

All bodies at temperature above absolute zero emit thermal radiation. Radiation is a volumetric phenomenon in the sense that all solids, liquids, and gases emit, absorb, or transmit radiation to varying degrees. The maximum rate of radiation follows the Stefan–Boltzmann Law for a blackbody, i.e.,

$$Q_{emit\ max}' = \sigma\, A_s\, T_s^4 = E_b',\qquad(5.18)$$

where $\sigma(= 5.669 \times 10^{-8}\ \text{W/(m}^2 \cdot \text{K}^4))$ is the Stefan–Boltzmann constant, T_s is in K (Kelvin) or °R (Rankine), and E_b' is the total blackbody emissive power. Figure 5.8 illustrates such an ideal surface (blackbody) at 10 K, and yes, even at the frostily frigid temperature of 10 K, it still gives out thermal energy!

Gray Surfaces   Real surfaces emit less radiation than ideal "black" ones. Emissivity, ε, is defined as

$$\varepsilon = E'/E_b',\qquad(5.19)$$

where E' is the emissive power of a real body and E_b' is that of a blackbody. Surfaces for which the emissivity is constant are called gray surfaces. With the introduction of ε to account for the non-ideality, i.e., non-black-body real surfaces or bodies, the radiation emitted from a real surface,

$$Q_{emit}' = \varepsilon\, \sigma\, A_s\, T_s^4,\qquad(5.20)$$

where ε takes on a value between 0 and 1. The emissivity of most building materials is approximately 0.9, over the range of typical building temperatures.

Radiation Properties: Absorptivity, Transmissivity, and Reflectivity   In addition to emissivity, ε, absorptivity, α_{rad}, transmissivity, τ_{rad}, and reflectivity, ρ_{rad}, also affect the rate of radiation heat transfer. Conservation of energy says that the

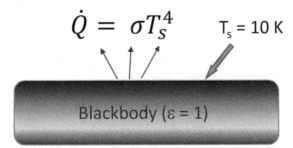

Figure 5.8.   A blackbody emitting thermal energy even at a frigid temperature of 10 K (created by D. Ting).

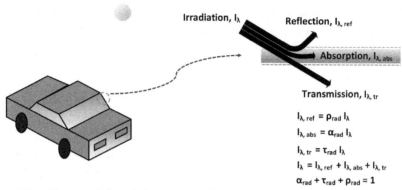

Figure 5.9. The total thermal energy radiated onto a surface is the sum of the absorbed, transmitted and reflected, $\alpha_{rad} + \tau_{rad} + \rho_{rad} = 1$ (created by X. Wang, edited by D. Ting).

amount of energy absorbed plus that transmitted through plus the portion reflected is 100% of the energy radiated on a material surface, i.e.,

$$\alpha_{rad} + \tau_{rad} + \rho_{rad} = 1. \qquad (5.21)$$

This is sketched in Fig. 5.9, noting that values for each of these falls between 0 and 1. Although Eq. 5.21 is derived for a single wavelength, it is valid for piecewise gray surfaces, if the wavelength range over which the three properties are calculated, is the same.

With absorptivity designating the fraction of the radiation energy incident on a surface that is absorbed, in this manner,

$$0 \le \alpha_{rad} \le 1, \qquad (5.22)$$

where $\alpha_{abs} = 1$ for a blackbody, which is both a perfect absorber and a perfect emitter. Kirchhoff's identity states that absorptivity and emissivity for gray surfaces (if the surface or the irradiation striking the surface is diffusive) are equal, i.e.,

$$\varepsilon = \alpha_{rad}. \qquad (5.23)$$

This expression is also valid for non-gray surfaces at a given wavelength. The rate of absorption, as illustrated in Fig. 5.10,

$$Q_{absorbed}' = \alpha_{rad}\, Q_{incident}'. \qquad (5.24)$$

Figure 5.11 shows the exchange of radiation between a surface and its surroundings. The net rate of radiation,

$$Q_{rad}' = \varepsilon \sigma A_s \left(T_s^4 - T_{sur}^4\right). \qquad (5.25)$$

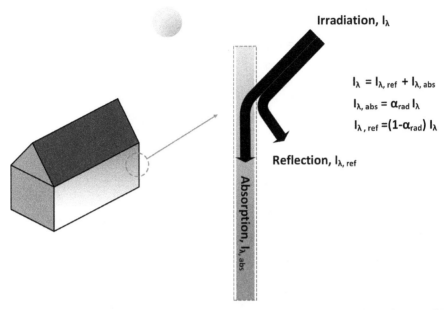

Figure 5.10.   Absorption and reflection of radiation onto a surface (created by X. Wang).

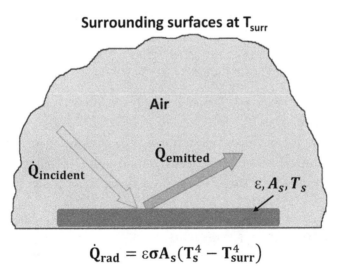

$$\dot{Q}_{rad} = \varepsilon\sigma A_s(T_s^4 - T_{surr}^4)$$

Figure 5.11.   Exchange of radiation between a surface and its surroundings (created by S.K. Mohanakrishnan, edited by D. Ting).

Note that the surrounding surface is at T_{sur}, in absolute units, i.e., either K or °R. Over a narrow (10~20 K) range of temperature, we may invoke the approximation,

$$Q_{rad}{}' = h_{rad}\, A_s\, (T_s - T_{sur}),\tag{5.26}$$

where the radiation heat transfer coefficient,

$$h_{rad} = \varepsilon\sigma\left(T_s + T_{sur}\right)\left(T_s^2 + T_{sur}^2\right). \tag{5.27}$$

In terms of the radiation resistance,

$$Q_{rad}' = \left(T_s - T_{sur}\right)/R_{rad}, \tag{5.28}$$

where the radiation resistance,

$$R_{rad} = 1/\left(h_{rad}\,A_s\right). \tag{5.29}$$

Shape Factors Radiation shape factor, F_{1-2}, denotes the fraction of radiation leaving diffuse surface 1 that is intercepted (but not necessarily absorbed) by surface 2. The shape factor is strictly geometric and does not depend on surface properties such as emissivity or temperature. We can deduce the reciprocity relationship,

$$A_1\,F_{1-2} = A_2\,F_{2-1}. \tag{5.30}$$

The first law of thermodynamics states that the total amount of energy must be conserved. Accordingly, the sum of shape factors for a given surface must equal unity, i.e.,

$$F_{1-1} + F_{1-2} + F_{1-3} + \dots + F_{1-j} = 1. \tag{5.31}$$

Note that the shape factor F_{1-1} is non-zero only for concave surfaces, i.e., surfaces that can "see" themselves.

It is important to note that shape-factor algebra described here applies only for diffusely emitting and reflecting surfaces. It cannot be used for surfaces such as mirrors that reflect specularly. However, nearly all surfaces in buildings are diffusive.

EXAMPLE 5.4. RADIATION FROM A FIREPLACE

Given: An old fashion, all black (including a long cylindrical section of the chimney, which is exposed in the room) wood fireplace is being replaced with a new, silver color natural gas one.

Find: What can you say about the efficiency, as far as keeping the room warm, of the new fireplace as compared to the old one? Why?

Solution:

The new fireplace is less efficient than the old one as far as effectively radiating the heat to occupants and indoor surfaces. This is because silver color tends to be shinny and has drastically lower emissivity than the good old dull black fireplace enclosure and chimney.

We are thus cautioned that not everything new is better than its older counterpart.

5.2 ABOVE-GRADE BUILDING HEAT TRANSFER

For engineering human thermal comfort, it is essential to have a good understanding of the steady-state heat transfer through the building envelopes. Walls, roofs, windows, and doors are the most common components making up the building envelope. The 2017 ASHRAE Handbook [ASHRAE, 2017], Chapter 26, Table 1, lists many building materials tested at 24°C. The heat transfer coefficients and the corresponding R-values include the combined effects of convection and radiation. Kreider and Rabl [1994] and Kuehn et al. [1998] are also very resourceful concerning both above-grade and below-grade building heat transfer.

5.2.1 Gas-Filled Cavity

Gas-filled cavities are quite common. Figure 5.12 is a schematic showing the heat transfer mechanisms involved. Between the two surfaces, one at T_H and the other at T_L separated by a gap of Δx, the heat transfer occurs via conduction, convection, and radiation. Typically, conduction heat transfer is very small compared to convection and radiation. Thus, the total heat flux,

$$q'' = q_{conv}'' + q_{rad}'' = (h_{conv} + h_{rad})(T_H - T_L). \qquad (5.32)$$

Since we are talking about cavities, the convective heat transfer mode is natural convection. The corresponding heat convection coefficient,

$h_{conv} = f$ (gas space position, heat flow direction, temperature difference, width of the gas space, thermal conductivity of the gas, k_{gas}); where k_{gas} is evaluated at T_{avg}.

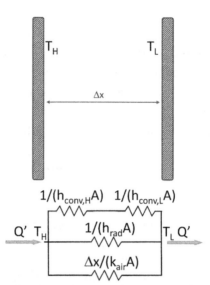

Figure 5.12.   Heat transfer through a gas-filled cavity (created by D. Ting).

It is clear that the practical way of dealing with radiation heat transfer in Eq. 5.32 is different from its fundamental format, as expressed by Eq. 5.25. The "practical" radiation heat transfer coefficient may simply be cast as,

$$h_{rad} \approx 4\sigma\, T_{avg}^3 / (1/\varepsilon_H + 1/\varepsilon_L - 1) \approx 4\varepsilon_{eff}\, \sigma\, T_{avg}^3, \qquad (5.33)$$

an alternate format to Eq. 5.27. The effective emittance,

$$\varepsilon_{eff} \equiv 1/(1/\varepsilon_H + 1/\varepsilon_L - 1). \qquad (5.34)$$

Two main assumptions have been invoked in dealing with radiation in the above manner. First, the gap spacing is small compared to the dimensions of the two surfaces. Hence, the view factor is roughly unity, and the surface areas are equal. Second, the temperature difference involved is about 10–20°C, and the temperatures are typically between −20 and 20°C. Common values of thermal resistance of plane air spaces can be found in Table 3, Chapter 26, of the 2017 ASHRAE Handbook [ASHRAE, 2017].

EXAMPLE 5.5. INCREASE THERMAL INSULATION OF AN AIR-GAP WALL

Given: An HVAC engineer wishes to evaluate the R-value of a 2 in wide air gap in a 100 ft² wall. For lowering the radiative heat transfer, she proposes lining the cavity's inner and outer surfaces with highly reflecting aluminum foil with $\varepsilon = 0.05$. On a typical winter day, the walls facing the gap are at 45 and 55°F. The convection coefficient at the inner surface of both walls is 0.32 Btu/(h·ft²·°F).

Find: The R-value of the cavity. The R-value without the aluminum foil, assume $\varepsilon = 0.9$.

Solution:

From Eq. 5.33,

$$h_{rad} \approx 4\,\varepsilon_{eff}\,\sigma\,T_{avg}^3.$$

where the effective emittance,

$$\varepsilon_{eff} \equiv 1/\left(1/\varepsilon_H + 1/\varepsilon_L - 1\right).$$

Substituting for $\varepsilon_H = \varepsilon_L = 0.05$ gives $\varepsilon_{eff} = 0.02564$, and therefore,

$$h_{rad} \approx 4\,(0.02564)\,5.669 \times 10^{-8}\,\text{W}/\left(\text{m}^2 \cdot \text{K}^4\right)(283\,\text{K})^3 = 0.1318\,\text{W}/\left(\text{m}^2 \cdot \text{K}\right).$$

The corresponding thermal resistance,

$$R = 1/h_{rad} = 7.588\,\text{m}^2 \cdot \text{K/W}.$$

Recall from Eq. 5.8 that the R-value is the unit thermal resistance,

$$R_{th} = AR = 7.588\,\text{m}^2 \cdot \text{K/W} \times 9.29\,\text{m}^2 = 70.5\,\text{K/W}.$$

If $\varepsilon_H = \varepsilon_L = 0.9$, then $\varepsilon_{eff} = 0.8182$, and hence,

$$h_{rad} \approx 4\,(0.8182)\,5.669 \times 10^{-8}\,\text{W}/\left(\text{m}^2 \cdot \text{K}^4\right)(283\,\text{K})^3 = 4.205\,\text{W}/\left(\text{m}^2 \cdot \text{K}\right).$$

The corresponding thermal resistance,

$$R = 1/h_{rad} = 0.2378\,\text{m}^2 \cdot \text{K/W}.$$

Thus,

$$R_{th} = 0.2378\,\text{m}^2 \cdot \text{K/W} \times 9.29\,\text{m}^2 = 2.2\,\text{K/W}.$$

It is clear that the aluminum foil is effective in increasing the R-value by more than 30 times!

5.2.2 Heat Transfer through Multi-Layered Structures

Figure 5.13 is a schematic of one-dimensional heat transfer through a multi-layered wall. During the heating season, the heat from the warm indoor air is convected to the inner surface of the indoor wall. It is then conducted through the series of walls, until it is finally convected to the cold outdoor air. The total resistance between the indoor and outdoor air,

$$R_{tot} = R_i + R_1 + R_2 + ... R_o = R_i + \sum R_j + R_o. \tag{5.35}$$

In terms of the heat transfer rate per unit area of wall, the heat flux,

$$q'' = (T_i - T_o)/R_{tot} = U(T_i - T_o). \tag{5.36}$$

We note that the U-value, $U = 1 / R_{tot}$. With wood studs in frame walls and roofs, concrete webs in concrete blocks, and metal ties, etc., in insulated wall panels, the heat transfer in reality is actually two or three dimensional. For all practical purposes, however, both the parallel-path and the isothermal-plane approximations can provide the needed accuracy.

Let us consider a simplified wood-frame wall section depicted in Fig. 5.14. In the winter time, some amount of the thermal system encompassed by the heated air indoor makes its way down the temperature gradient, into the outdoors. Imagine we ride on Ms. Frizzle's Magic School Bus, sliding down the temperature slope. The first thermal resistance is the convection heat transfer resistance between the indoor air and the inner wall, which is typically drywall. The next thermal resistance is the conduction resistance of the first layer of the building envelope, the drywall. Between the inner wall (drywall) and the outside wall, typically brick or concrete, is a two-path passage. The Magic School Bus can either go through the

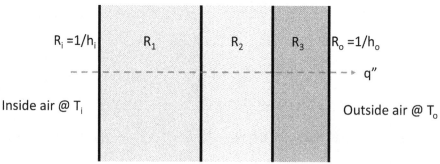

Figure 5.13.   One-dimensional heat transfer through a multilayered wall (created by D. Ting).

easier, less thermally resistive path, which is typically the studs used for supporting, or, the notably more resistive "pink" insulating material. There are two common ways to deal with this, i.e., (a) the parallel-path method and (b) the isothermal-plane method.

Parallel-Path Method The parallel-path method assumes that the portion of the heat transmitted through the less resistive path involving the relatively higher conductive stud or other materials remains parallel from the indoor wall to the outside surface. Imagine a lineup of Magic School Buses, each carrying Q amount of heat, start before the inside wall of Fig. 5.14. If Buses 1 and 2 lined up above the top stud, Buses 3–6 between the horizontal lines bounding the top stud, Buses 7–10 between the two studs, Buses 11–14 through the lower stud, and Buses 15 and 16 below the bottom stud. In this way, we would expect Buses 3–6 and 11–14, which encounter a section of the lower resistance stud path, to travel faster than Buses 1, 2, 7–10, 15 and 16. This is portrayed in Fig. 5.15, where Q_{stud}' is the heat transfer rate along the parallel horizontal path, which consists of a section of lower resistance stud, and the lower heat transfer rate, Q_{insul}', through the parallel path, which includes a section of high-resistive insulation path. The total heat transfer rate,

$$Q_{tot}' = Q_{stud}' + Q_b' = [A_{stud} (T_i - T_o)]/R_{tot, stud} + [A_b (T_i - T_o)]/R_{tot,b}, \quad (5.37)$$

where subscript "b" denotes the board-insulation pathway. The total average thermal resistance,

$$A_{tot}/R_{tot, avg} = A_{stud}/R_{tot, stud} + A_b/R_{tot, b}. \quad (5.38)$$

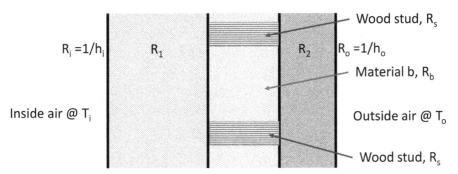

- Area Ratio, $A_{s-b} = A_s/(A_s + A_b)$
- $Q_{tot}' = A_{tot} (T_i - T_o) / R_{tot,avg} = U_{avg} A_{tot} (T_i - T_o)$.
- The average U-value, $U_{avg} = 1 / R_{tot,avg}$.

Figure 5.14.   A schematic of a typical wood-frame wall section (created by D. Ting).

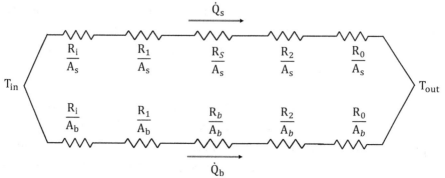

Figure 5.15. Equivalent-circuit diagram for the parallel-path method (created by R. Reeghesh).

The average total conductance,

$$\frac{1}{R_{tot,\,avg}} = \frac{A_{stud}}{A_{tot}} \frac{1}{R_{tot,\,stud}} + \frac{A_b}{A_{tot}} \frac{1}{R_{tot,\,b}}. \tag{5.39}$$

The underlying assumption here is that all Magic School Buses stay on their course, even when they are confronted with much slower traffic. In reality, however, some of these buses, especially the ones just next to a less congested passage, would take advantage of the speedway, rendering the heat transfer two or three dimensional. We note that for the parallel circuit associated with the parallel-path method, it is more convenient to use the U-value, as

$$U_{avg} A_{avg} = \sum \left(U_j A_j \right). \tag{5.40}$$

Isothermal-Plane Method Alternatively, we may treat the multi-layered heat transfer problem using the isothermal-plane method. For this method, it is assumed that the lateral heat transfer is excellent, giving a constant temperature along each lateral interface plane, i.e.,

$$T = T(x) \text{ only, i.e., } \neq f(y). \tag{5.41}$$

We may envision this using the Magic School Buses illustration again. For this isothermal-plane method, all the buses are bolted together in a line parallel to the plane (wall). As such, the slower buses passing through the more resistive portions of the road are dragged forward by the faster moving ones on the less-resistive path. As the faster buses drag the slower ones, they slow down. Hence, all buses move across the wall in unison, and therefore, the temperature is uniform over any plane.

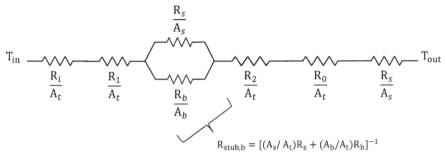

Figure 5.16. Equivalent-circuit diagram for the isothermal-plane method (created by R. Reeghesh and Y. Yang).

A thermal circuit is shown in Fig. 5.16. According to this thermal circuit, the total average resistance,

$$R_{tot,\,avg} = R_{in} + R_1 + R_{stud,\,b} + R_2 + R_{out}, \qquad (5.42)$$

where

$$A_{tot}/R_{stud,\,b} = A_{stud}/R_{stud} + A_b/R_b, \qquad (5.43))$$

i.e.,

$$R_{stud,\,b} = \left(\frac{A_s}{A_b}R_s + \frac{A_b}{A_t}R_b\right)^{-1}. \qquad (5.44)$$

The total heat transfer rate,

$$Q_{tot}{}' = A_{tot}\,(T_i - T_o)/R_{tot,\,avg} = U_{avg}\,A_{tot}\,(T_i - T_o). \qquad (5.45)$$

It is clear from Eq. 5.42 that it is easier to use the R-value when applying the isothermal-plane method.

It is worth stressing that either the parallel-path method or the isothermal-plane method is good enough for most HVAC applications. One could check this by applying both methods and comparing the corresponding results. In special cases in which there is a significant difference, and a better accuracy is needed, a computational fluid dynamic simulation may be called upon.

5.2.3 Heat Transfer through Windows

Occupied spaces such as residences, offices, and schools without adequate fenestration would resemble dungeons, negatively affecting and slowly breaking the occupants. Thankfully, the spirit-lifting building facade is becoming ever more popular. Much details on fenestration are disclosed in Chapter 15 of the 2017 ASHRAE Handbook [ASHRAE, 2017]. According to the purpose of this book, the basic elements underlying heat transfer through windows are explicated in a reader-friendly way; the readers are reminded to refer to the practical standard, ASHRAE Handbook, as needed.

For a typical window sketched in Fig. 5.17, the rate of heat transfer through it, when neglecting sunlight, solar irradiance, air infiltration and exfiltration, moisture condensation, etc., can be expressed as

$$Q_{overall}{'} = U_o A_o (T_i - T_o), \tag{5.46}$$

where U_o is the overall heat transfer coefficient, i.e., the U-value, and A_o represents the combined glazing plus frame area. The overall U-value,

$$U_o = \left(U_{cg} A_{cg} + U_{eg} A_{eg} + U_f A_f\right) / \left(A_{cg} + A_{eg} + A_f\right), \tag{5.47}$$

where subscripts cg, eg, and f refer to the center-of-glass, edge-of-glass, and frame, respectively.

As aluminum and its alloys are practicable materials for the frame, means have to be implemented to overcome the shortcoming of their high conductivity. Thermal breaks made of very high thermally resistive materials are inserted between the outside and inside metal frames (see Fig. 5.18). In this way, the heat carried by the Magic School Buses is literally halted at these thermal breaks.

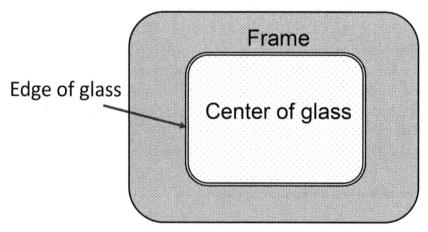

Figure 5.17.   Typical components making up a window (created by D. Ting).

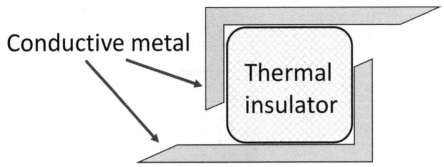

Figure 5.18. A thermal break for reducing heat loss through a highly-conductive (metallic) window frame (created by D. Ting).

5.3 BELOW-GRADE BUILDING HEAT TRANSFER

Basements in Canada and other parts of the world are an integral part of residential living. Some basements are well furnished and utilized, mostly for entertainment, including serving as an indoor playground and/or as extra guest rooms. More commonly, a part of the basement is used to house the HVAC systems and the laundry washing and drying machines. Cooler basements during the summer months do not typically invoke any notable cooling concern. In the winter time, on the other hand, there can be notable heat loss through the basement walls and floor. Szydlowski and Kuehn [1981] showed that the conventional simplified one-dimensional steady-state analysis can provide good estimates for design heat losses. Understandably, many software programs are available to provide a more detailed analysis. Even so, it is critical for HVAC engineers to have a proper appreciation of the underlying physics revealed by the simple steady-state analysis.

5.3.1 Through Basement Walls

Ground coupling is heat transfer between the basement or floor slab of a building and the earth. Although several methods for calculating ground coupling have been suggested, the corresponding values of heat transfer rates can disagree significantly sometimes [Claridge, 1986]. Field measurement of an uninsulated basement by Latta and Boileau [1969] showed that the isotherms near the wall are roughly radial lines centered at the intersection of the grade and the wall. In plain English, the two-dimensional heat flow paths from a basement wall, when viewed in cross section, are circular arcs centered at the intersection of the wall and the earth's surface (see Fig. 5.19).

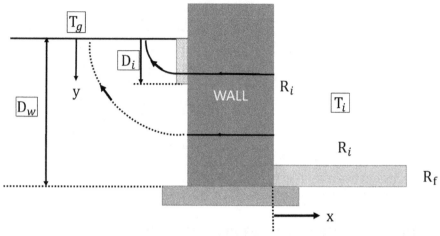

Figure 5.19. Heat transfer from below-grade (basement) to outdoor; see, e.g., Figs. 12–14, page 18.39 of 2017 ASHRAE Handbook (ASHRAE, 2017) (created by R. Reeghesh, edited by D. Ting).

Note from Fig. 5.19 that the outdoor air is colder than the ground. Therefore, the heat transfer is from the warmer basement air, through the basement wall, radially through the soil, to the ground surface, which is at the ambient temperature. Following Chapter 18 of the 2017 ASHRAE Handbook [ASHRAE, 2017], the approximate steady-state heat loss from the basement to the ground surface may be expressed as [Latta & Boileau, 1969],

$$Q' = U_{avg} A \left(T_i - T_g\right), \tag{5.48}$$

where U_{avg} is the average U-factor, T_i is the basement temperature, and the design ground temperature, the soil temperature at 10 cm (4 in) soil depth,

$$T_g = T_{g, avg} - T_A \tag{5.49}$$

Here, $T_{g,avg}$ is the mean deep ground temperature, which is estimated from the annual average air temperature or from well-water temperatures (see Fig. 18 of Chapter 34, 2011 ASHRAE Handbook—HVAC Applications). T_A is the amplitude of the ground surface temperature or deviation from the deep soil temperature (see Fig. 13, Chapter 18 of the 2017 ASHRAE Handbook [ASHRAE, 2017]).

It is clear that, because of the progressively longer heat path farther into the ground, there is little incentive to insulate the basement wall all the way down. With finite or partial insulation, the heat flow paths take shapes somewhere between circular and vertical lines.

5.4 MOISTURE TRANSPORT

The diffusion of water vapor is a primary concern in HVAC. This is particularly the case during the cold winter months, where the transport of water vapor from the warm indoor environment through the progressively colder building envelope can lead to condensation and freezing. Accordingly, vapor retarders and/or barriers are placed on the warm side to mitigate this. Very much like natural (biological) flatulence, water vapor diffuses from higher to lower concentration regions. The concentration of water vapor corresponds to its partial pressure. In other words, water vapor diffuses from higher to lower vapor pressure regions, i.e.,

$$m_w' \propto A \, dP_w/dx. \tag{5.50}$$

Replacing the proportionality with a constant, we have the water vapor flux,

$$m_w'/A = -\mu \, dP_w/dx, \tag{5.51}$$

where μ is the permeability of water vapor of the building material. The negative sign implies that the diffusion takes place in the direction of decreasing vapor pressure.

The value of permeability, μ, is dependent on the material, pressure, and humidity, among other factors. The permeability per unit depth is called moisture transport conductance or permeance,

$$M = \mu/L. \tag{5.52}$$

The water vapor flux can be expressed in terms of this permeance as

$$m_w'/A = M \, \Delta P_w. \tag{5.53}$$

Alternatively, we can use water vapor-flow resistance,

$$Z = 1/M. \tag{5.54}$$

Typical values of these parameters for HVAC applications can be found in the ASHRAE Handbook (ASHRAE, 2017).

Figure 5.20 depicts an ideal, one-dimensional heat and moisture transfer through a multi-layer wall. The heat flux,

$$q'' = (T_{warm} - T_x)/R_{w-x} = (T_{warm} - T_{cold})/R_{tot}. \tag{5.55}$$

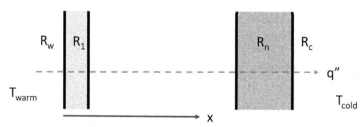

Figure 5.20. One-dimensional heat and moisture transfer (created by D. Ting)

where R_{w-x} signifies the thermal resistance from the warm indoor (air) to location x in the multi-layer wall. Under steady-state conditions, this heat flux is a constant. On that grounds, the temperature at location x can be deduced from

$$T_x = T_{warm} - (R_{w-x}/R_{tot})(T_{warm} - T_{cold}). \qquad (5.56)$$

Under steady-state conditions, the continuous no-condensation water vapor flux,

$$m_w'/A = (P_{w,warm} - P_{w,cold})/Z_{tot} \qquad (5.57)$$

or,

$$m_w'/A = (P_{w,\,warm} - P_{w,\,x})/Z_{w-x} \qquad (5.58)$$

or, from location x to the cold outdoor (air),

$$m_w'/A = (P_{w,\,x} - P_{w,\,cold})/Z_{x-c} \qquad (5.59)$$

from which, we can write,

$$(P_{w,\,warm} - P_{w,\,cold})/Z_{tot} = (P_{w,\,warm} - P_{w,\,x})/Z_{w-x} \qquad (5.60)$$

or,

$$(P_{w,\,warm} - P_{w,\,cold})/Z_{tot} = (P_{w,\,x} - P_{w,\,cold})/Z_{x-c}. \qquad (5.61)$$

The total water-vapor resistance,

$$Z_{tot} = Z_{x-c}(P_{w,\,warm} - P_{w,\,cold})/(P_{w,\,x} - P_{w,\,cold}). \qquad (5.62)$$

Combining the first two expressions for water vapor flux gives

$$P_{w,\,x} = P_{w,\,warm} - (P_{w,\,warm} - P_{w,\,cold})Z_{w-x}/Z_{tot}. \qquad (5.63)$$

For all locations, $P_{w,x} < P_{w,sat}$ is a necessary condition to avoid condensation. To prevent condensation, the vapor retarder with vapor resistance,

$$Z_{vr} > Z_{x-c}(P_{w,\,warm} - P_{w,\,cold})/(P_{w,\,x} - P_{w,\,cold}) - Z_{tot-vr} \qquad (5.64)$$

on the warm side, where Z_{tot-vr} is the sum of all the vapor flow resistance excluding that of the vapor retarder. In practice, it is generally desirable to locate the vapor retarder as close to the warm side as possible.

PROBLEMS

Problem 5.1

Figure 5.21 shows a building wall. The inner wall is made of 2-cm thick gypsum board with a thermal conductivity of 0.2 W/m·K. Steel studs with a thermal conductivity of 40 W/m·K are used to secure the inner wall to the outer wall. They are 10 cm long (the distance between the inner and outer walls) with a 5 cm by 5 cm cross section. The steel studs are placed 0.5 m apart, both horizontally and vertically. The 25-cm thick outer wall is made of brick with thermal conductivity of 1 W/m·K. If the temperature of the inner surface of the gypsum is at 18°C and that of the outer surface of the brick is at 35°C, what is the heat flux through the wall?

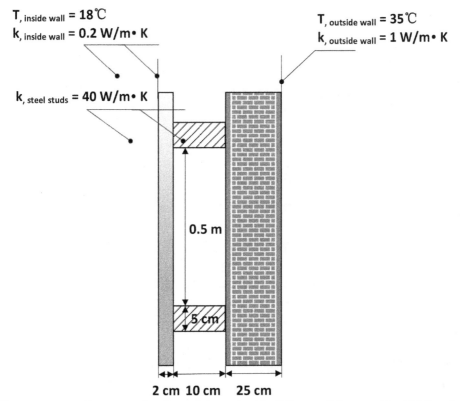

Figure 5.21. Heat transfer through a typical building wall (created by X. Wang).

Problem 5.2

The outside wall of a home consists of a 10-cm layer of common brick (k = 0.7 W/m/K), a 15-cm layer of fiberglass insulation (k = 0.038 W/m/K), and a 1-cm layer of gypsum board (k = 0.4 W/m/K). What is the overall R-value? What is the heat flux through the wall if $T_i = 22°C$ and $T_o = 5°C$?

Problem 5.3

For structural reasons, the wall in Problem 5.2 must have studs placed every 60 cm. The studs are fabricated from wood (k = 0.1 W/m/K) and are 5 cm wide and 15 cm deep. What are the corresponding R-value and heat flux?

Problem 5.4

The exterior wall of a brick building is constructed of 27-cm thick face brick (conductivity k = 1.3 W/m·K, emissivity ε = 0.9), a 9-cm air gap, and 27-cm thick common brick (conductivity k = 0.7 W/m·K, emissivity ε = 0.9). During a winter night, the outer surface is –32°C and the inner surface is 8°C. Find the heat flux, q" through the wall.

Problem 5.5

Choose the most appropriate answer concerning a gas-filled cavity. With an air gap of 10 cm, one surface at 18°C and the other at 7°C, which of the following is likely going to allow the highest heat transfer rate?

A) The wall-cavity-wall is vertical
B) The wall-cavity-wall is at a 20-degree tilt, with the warmer wall above the cooler one
C) The wall-cavity-wall is at a 20-degree tilt, with the cooler wall above the warmer one
D) The wall-cavity-wall is at a 10-degree tilt, with the warmer wall above the cooler one
E) The wall-cavity-wall is at a 10-degree tilt, with the cooler wall above the warmer one

Problem 5.6

A wall, from inside to outside, consists of 9.5-cm thick gypsum board, fiber glass, 12.7-mm thick sheathing next to the insulation and wood framing (wood framing makes up 20% area), 38-cm thick polystyrene, 10-cm thick brick. The indoor air is at 20°C, and the corresponding heat transfer coefficient is 8 W/m²·K. The outdoor is at −10°C with a wind speed of 13 km/h, the corresponding heat transfer coefficient is 25 W/m²·K.

Layer, n	Actual area (m²)	k_n (W/m·K)	Thickness (mm)	R_n (m²K/W)
Indoor air film	1			0.125
Gypsum	1	0.16	9.5	0.059
Fiber glass	0.8	0.04	140	3.5
Wood frame	0.2	0.15	140	0.93
Sheathing	1	0.055	12.7	0.23
Polystyrene	1	0.036	38	1.06
Brick	1	0.9	100	0.11
Outdoor air film				0.04

Find the average heat transfer rate through the wall based on (a) the parallel-path method (b) the isothermal-plane method.

Problem 5.7

A 8-ft high and 20-ft deep unheated garage has a 10-ft span exterior wall with U = 0.2 Btu/hr·ft²·°F. The other three walls (U = 0.05 Btu/hr·ft²·°F) separate it from the heated indoor with an inner surface at 72°F. The ceiling plus attic plus roof has a U-value of 0.7 Btu/hr·ft²·°F. The outside air is at −5°F. Find the temperature of the garage.

Problem 5.8

A 10,000 ft² flat roof has its lower surface at 120°F. The ceiling is suspended 2 ft below with its top surface at 85°F. The emissivity of all surfaces may be approximated to be 0.9. Find the shape factor, heat transfer rate, and the relative magnitude of convection.

Problem 5.9

A flat roof consists of three layers of materials. The top layer is made of plywood and shingles with a total of R3 in US Customary Units. The middle layer consists of sheets of 15-cm thick fiberglass. The lowest layer is made of 1.2-cm thick drywall. On a particular day, the indoor and outdoor temperatures are 68°F and 10°F, respectively. Find the location at which the temperature is 32°F. Where should the vapor retarder be placed?

Problem 5.10

A 100 m^2 ceiling with an R-value of 5 K·m^2/W has a heat transfer coefficient of 8 W/(m^2·K) for both the upper and lower surfaces. The 144 m^2 roof has a R-value of 0.5 K·m^2/W, the heat transfer coefficient on the lower surface is 8 W/(m^2·K), and the heat transfer of the upper surface is 24 W/(m^2·K). On a winter night, the outdoor temperature is at −25°C, whereas the indoor temperature is maintained at 18°C. What is the rate of heat transfer through the ceiling? What is the attic air temperature?

Problem 5.11

A 5 m by 7 m radiant floor has a surface temperature of 28°C. The 2.5-m high walls and the ceiling are at 17°C. What is the heat transfer rate from the radiant floor if all surfaces have an emissivity of 0.9?

Problem 5.12

Choose the most appropriate answer concerning winter heat loss through basement wall. Assuming that the conductivity of the soil is a constant, how does the thermal resistance increase with depth, y (into the soil)?

A) The thermal resistance, R, increases with depth, y.
B) The thermal resistance, R, increases with ½ pi times depth, ½πy.
C) The thermal resistance, R, increases with pi times depth, πy.
D) The thermal resistance, R, increases with the square root of depth, √y.
E) The thermal resistance, R, decreases with the square root of depth, √y.
F) The thermal resistance, R, increases with the square of depth, y^2.

REFERENCES

American Society of Heating, Refrigeration and Air Conditioning Engineers (ASHRAE), *2017 ASHRAE Handbook: Fundamentals*, SI Edition, ASHRAE, Atlanta, 2017.

T.L. Bergman, A.S. Lavine, F.P. Incropera, D.P. Dewitt, *Introduction to Heat Transfer*, 6th ed., Wiley, Hoboken, 2011.

Y.A. Çengel, A.J. Ghakar, *Heat and Mass Transfer: Fundamentals & Applications*, 5th ed., McGraw-Hill, New York, 2015.

D. Claridge, "Building to ground heat transfer," Proceedings of American Solar Energy Society Conference, Boulder, Colorado, pp. 144–154, 1986.

J.F. Kreider, A. Rabl, *Heating and Cooling of Buildings: Design for Efficiency*, McGraw-Hill, New York, 1994.

T.H. Kuehn, J.W. Ramsey, J.L. Threlkeld, *Thermal Environmental Engineering*, 3rd ed., Prentice-Hall, Upper Saddle River, 1998.

J.K. Latta, G.G. Boileau, "Heat losses from house basement," Canadian Building, 19(10): 39–42, 1969.

R.F. Szydlowski, T.H. Kuehn, "Analysis of transient heat loss in earth-sheltered structures," Underground Space, 5: 237–246, 1981.

CHAPTER 6

Heating in the Winter

"To appreciate the beauty of a snowflake it is necessary to stand out in the cold."

–Aristotle

CHAPTER OBJECTIVES

- Appreciate the need for space heating and cooling.
- Be aware of typical outdoor and indoor design conditions.
- Focus on winter heating, without the complications imposed by solar radiation.
- Estimate building air exchange.
- Understand the "air change" and "crack" methods in evaluating air exchange.
- Resolve the wind, stack, and ventilation (pressurization) effects on air exchange.

Nomenclature

A	Area; A_{leak} the effective air leakage area
A_{CH}	Air change rate
a_s	An empirical coefficient associated with the stack effect on air leakage
a_w	An empirical coefficient associated with the wind effect on air leakage
C	Empirical coefficient; C_P is pressure coefficient, C_s is thermal draft coefficient
c	Flow coefficient
g	Gravity

(Continued)

H	Height; ΔH is height difference
h	Heat transfer coefficient; h_{summer} is the heat transfer coefficient for an exterior surface in the summer, h_{winter} is the heat transfer coefficient for an exterior surface in the winter
HVAC	Heating, ventilation, and air conditioning
LED	Light-emitting diode
n	Exponent
P	Pressure; P_i is the indoor pressure, P_o is the outdoor pressure, ΔP is the pressure difference, ΔP_p is the pressure difference due to building pressurization, ΔP_s is the pressure difference due to stack effect, ΔP_w is the wind-induced pressure difference, $\Delta P_{w,m}$ is the maximum pressure difference induced by the wind
R	Gas constant; R_a is the gas constant of air, R_i is the gas constant of indoor air, R_o is the gas constant of outdoor air
T	Temperature; T_i is the indoor temperature, T_o is the outdoor temperature, ΔT is temperature difference
V	Velocity or speed; V_{summer} is the typical wind speed in the summer, V_w is the wind speed, V_{winter} is the typical wind speed in the winter

Greek and Other Symbols

ρ	Density; ρ_i is the indoor air density, ρ_o is the outdoor air density
$\forall$	Volume; $\forall'$ is volume flow rate

6.1 THE NEED FOR SPACE HEATING AND COOLING

Most of the time, some amount of thermal energy must be supplied to, or removed from, a building in order to maintain comfortable indoor conditions, i.e., ensuring human thermal comfort. The amount of thermal energy required to keep a building thermally comfortable to occupy is called the heating load. On the other hand, the quantity of thermal energy that needs to be removed during the summer months is the cooling load.

In our efforts to save energy, during the oil crisis in the 1970s, in particular, many buildings were well-sealed and air-tight [de Oliveira Fernandes, 1994]. Shortly after, "sick building syndrome" was coined to describe the unwellness associated with staying too long in a tightly sealed building, which not only lacks fresh air, but also contains too much of many indoor emissions from various sources, including building materials and human natural venting such as exhalation, expiration, and expulsion of gas.

We thus see two necessities going hand-in-hand, i.e., the need for heating or cooling and a minimum exchange of indoor with outdoor air. Concerning the minimum air exchange, a part of it is driven by nature, outdoor wind, and temperature. As mentioned in the previous paragraph, this is typically inadequate for a properly sealed building. Therefore, measures must be appropriately engineered into the HVAC system. More on air exchange will be discussed. It is appreciated that the required minimum air exchange furthers the need for heating and cooling, especially when the weather or outdoor condition deviates from the thermally comfortable one.

In the simplest form, the only big question that needs to be answered during the design and selection of the HVAC system for a building is the required capacities. In other words, the size of the heating system must be properly selected to ensure adequate heating during the cold winter, see Fig. 6.1. Similarly, the capacity of the

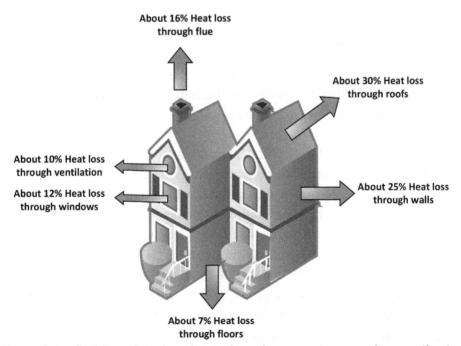

Figure 6.1.   Building winter heat loss; where the percentages are loose estimates (created by X. Wang).

cooling system must meet the heat removal rate required to keep the indoor habitable during the hot summer. The sizing of the heating and cooling units is still all that is needed for simple applications such as a typical residence.

For larger buildings, or in industrial settings that utilize large amounts of energy, the annual heating and cooling costs also come into the forefront. The selection of the HVAC system includes proper consideration of capital and operation costs. Life cycle analysis, payback time, etc., are typical cost analyses exploited in this case. The operation cost is composed primarily of the total annual energy consumption. As such, details, including the variation in the heating and cooling requirements throughout a typical year, are scrutinized for deducing unique measures to lower costs. This is called load calculation, where a careful accounting of all the thermal energy terms in a building is executed.

6.2 TYPES OF BUILDING HEAT LOSS AND GAIN

In general, there are three main types of heat loss or gain:

1. Heat transmission through the building envelope. This takes place in the form of convection and radiation followed by conduction, and subsequently, convection and radiation, through the walls, windows, roof, and doors. The steady-state heat transfer expounded in Chapter 5 is generally adequate as a first estimate for all practical purposes. For reducing the heating and cooling loads, the general approach is to maximize the solar heat gain during the winter, minimize it during the summer, and increase the building envelope insulation.

2. Infiltration–exfiltration. Infiltration–exfiltration is the uncontrolled airflow through those inevitable cracks and openings of a building. In a way, infiltration, which is countered and balanced by exfiltration, serves as a safeguard in mitigating sick-building syndrome. Unfortunately, natural infiltration–exfiltration is rather unevenly distributed, leaving certain occupied places, especially the inner spaces that are not in contact with an external wall, below the desirable condition. Worse still, these inner spaces away from the building envelope may end up lacking fresh air in the absence of mechanical ventilation. Also, uncontrolled natural infiltration wastes energy. In other words, air exchange in excess of the necessary implies energy wastage, unless the outdoor air is at the ideal supply air condition and that minimal effort is needed

to augment the air exchange.[1] In short, controlled ventilation is the way forward. When this is complemented with a heat recovery system, it can also lead to significant energy savings.

3. Solar radiation. The dynamic nature of solar radiation imposes a considerable amount of complication to the required cooling load estimation. Heating load, to a great extent, is not sensitive to the dynamic aspect of solar radiation. For this reason, we cover the heating load first in this chapter. The solar-dependent cooling load is discussed after a dedicated chapter detailing solar radiation.

Internal gains from occupants, appliances, etc., are typically small. One exception is lighting, which can amount to some moderate to notable amount of internal heat gain. This is why much recent effort has been invested in replacing existing lights with significantly more energy-efficient lights. Over the years, incandescent lights, which are more-or-less extinct, have been phased out by florescent lights. The relatively new florescent lights are being replaced by light-emitting diode (LED) lights. Nevertheless, energy-intensive spotlights are still quite common, awaiting replacement with LED ones.

6.3 WINTER VERSUS SUMMER DESIGN CONDITIONS

Typical peak loads need to be estimated for the proper sizing of the heating and cooling equipment. These are based on multi-year averages of the typical maximum hot and cold weather conditions. For efficient utilization of the HVAC equipment, designing for the most extreme weather on record is not practical. In plain English, aiming to cover record highs and lows will end up significantly over-sizing the heating and cooling systems. In general, it is acceptable for the occupants to put on some extra clothing, and/or resort to use some auxiliary heating, during those record-breaking winter nights. During those few days of high-summer records, extra showers along with fanning and/or dehumidification tend to enable the occupants to cope with the extra heat and/or humidity. Therefore, designing to around

---

[1]Any increase in fan power over fan usage period is added cost. Therefore, bringing in more than necessary fresh outdoor air, even if it is at the ideal supply air condition, into the building is not executed, except when it can be realized freely and easily such as by opening the windows.

a 95% multi-year averaged high and low will suffice. There are exceptions, however, seniors and the physically-ill in nursing homes and hospitals should not be subject to the same extent of adaptation adjustments. Consequently, tighter multi-year averages are employed. Just when you think that human life and well-being are most precious, those lifeless and, at times, priceless items in museums are considerably more pampered with extravagant HVAC systems, which guarantee close to 100% extreme weather coverage. Even the humidity has to be regulated within a very narrow bandwidth to safeguard these precious articles.

The thermal properties of building materials in practice are somewhat affected by the circumstances involved. As such, the R-values and/or U-values of construction materials are tested and graded under standard conditions. The surface of the material is typically at 21°C, whereas the surrounding air is at 15.5°C; see Table 24.1 of the 1997 ASHRAE Handbook [ASHRAE, 1997]. It is noted that the heat-transfer coefficients and corresponding R-values include the combined effects of convection and radiation.

Concerning the R-values, the convection heat transfer portion is primarily a function of the wind speed. Wind speed is very much a weather-dependent variable; the conventional rough estimates are [Kreider et al., 2002]:

$$V_{winter} = 6.7 \, \text{m/s} \tag{6.1}$$

and

$$V_{summer} = 3.4 \, \text{m/s.} \tag{6.2}$$

The corresponding heat transfer coefficient (radiation + convection) for the exterior (vertical) surfaces are

$$h_{winter} = 34.0 \, \text{W/} \left(\text{m}^2 \cdot \text{K} \right) \tag{6.3}$$

and

$$h_{summer} = 22.7 \, \text{W/} \left(\text{m}^2 \cdot \text{K} \right). \tag{6.4}$$

These are summarized in Table 6.1. Today, much better estimates can be obtained from the ever more detailed climate design information that has been compiled. One of the best resources is the 2017 ASHRAE Handbook [ASHRAE, 2017]. The companion CD-ROM contains data collected at 8118 locations in the United States, Canada, and around the world, an increase of 1675 locations from its previous edition. Data for T_{dry}, T_{wet}, wind speed with direction and frequency of occurrence,

Table 6.1. Approximate first estimates of seasonal wind speed and heat transfer coefficient.

Season	Wind speed (m/s)	Heat transfer coefficient [W/(m²·K)]
Summer	3.4	22.7
Winter	6.7	34.0

etc., are catalogued. The values of different averages corresponding to 0.4, 1, and 2 percentages of warm-season temperature and humidity and 99.6% and 99% for the cold-season are also given to ease design. Note that 99% coverage indicates that there will be less than 88 hours in a year in which the outdoor condition will be below or above that corresponding design value. For example, 99% minimum temperature implies that there will be less than 88 hours in a typical year in which the outdoor temperature will be below the tabulated value. Moving the percentage to 99.6% means that less than 35 hours of the 8760 hours in a year will have temperatures below the corresponding tabulated value.

EXAMPLE 6.1. ATMOSPHERIC CONVECTIVE COOLING OF SOLAR PANELS

Given: An array of solar panels measuring 10 m by 10 m is placed on a flat roof. The panels are heated by solar radiation so that they are 20°C warmer than the ambient air.

Find: The convective heat loss from the solar panels for (a) no wind, (b) summer design wind speed of 3.4 m/s, and (c) winter design wind speed of 6.7 m/s.

Solution:

Among the multitude of literature on convective heat transfer from solar collectors, the following three seem to be representatively sound for the problem at hand. The convection heat transfer coefficient, h, is in W/(m² · °C), and the wind speed, V, is in m/s.

   Cole and Sturrock [1977]:

$$h = 5.7\,V \tag{6.1.1}$$

   Test et al. [1981]:

$$h = 2.56\,V + 8.55 \tag{6.1.2}$$

(Continued)

Sharples and Charlesworth [1998]:

$$h = 3.3\,V + 6.5 \tag{6.1.3}$$

(a) No wind, $V = 0$, the corresponding values for h for the aforementioned equations are

$$h = 0,\ 8.55,\ 6.5\,W/\left(m^2 \cdot {}^\circ C\right)$$

This is natural convection, and we expect some finite heat convection. Hence, Cole and Sturrock equation is possibly not meant for very low wind speeds.

(b) $V = 3.4$ m/s, the corresponding values for h for the aforementioned equations are

$$h = 19.38,\ 17.25,\ 17.72\,W/\left(m^2 \cdot {}^\circ C\right)$$

(c) $V = 6.7$ m/s, the corresponding values for h for the aforementioned equations are

$$h = 38.19,\ 25.70,\ 28.61\,W/\left(m^2 \cdot {}^\circ C\right)$$

We see that the value varies quite significantly among the three expressions and also in comparison to those listed in Table 6.1; noting that those tabulated in Table 6.1 are primarily meant for vertical walls. With this backdrop, we appreciate that the estimate of a convection heat transfer coefficient is only good to approximately ±30%.

6.4 BUILDING AIR EXCHANGE

As mentioned earlier, when there is inadequate exchange between the indoor air and the outdoor air, the occupants can suffer from sick-building syndrome. Too much fresh air, on the other hand, implies energy wasting. In plain English, extra energy is required to condition the excess fresh air so that the indoor environment remains thermally comfortable. The most common measure for quantifying air exchange is air change per hour,

$$A_{CH} = \forall'/\forall, \tag{6.5}$$

where $\forall$ is the building air volume and $\forall'$ is the volumetric flow rate of outdoor air into the building. The building designer can estimate the air exchange rate based on data from similar buildings or via building modeling. The modeling approach can be more precise, but a fair amount of effort may be required.

As introduced earlier, the two mechanisms that contribute to the total air exchange are (a) infiltration–exfiltration and (b) ventilation. Infiltration–exfiltration is usually the main air exchange mechanism in residences and small business buildings. Older buildings typically have 1–2 air changes per hour (AC/h), which is more than enough for acceptable indoor air quality. The typical values for newer buildings are significantly lower at 0.3 to 0.7 AC/h. The air exchange rate is particularly influenced by the outdoor-to-indoor pressure and temperature differences. The portion attributed to mechanical ventilation is largely driven by the exhaust fans and also the openings of windows and doors. When operating a powerful kitchen exhaust fan, one should make sure that the incoming outdoor air comes through the desired openings; a window through which clean outdoor air is plentiful may need to be opened. For a small residence, leaving the recirculating fan on can improve indoor air quality by homogenizing localized poor air quality with the large volume of the total indoor air. Natural ventilation is mostly dictated by the height of the building, along with building openings, including windows and doors.

It is stated correctly in Chapter 16 of the 2017 ASHRAE Handbook—Fundamental [ASHRAE, 2017], the only reliable way to deduce the air exchange rate of an existing building is to measure it. Interested readers can refer to the authoritative handbook for details. Briefly, the air exchange rate can be assessed directly by means of a tracer gas. This direct estimation way of estimating the air leakage is called the **Air Change Method**. For example, sulfur hexafluoride, SF_6, an inert and harmless gas, which can be detected at concentrations above 1 ppb, has been successfully employed [Sherman et al., 1980; Harrje et al., 1981]. The building under study can be pressurized via a blower door where a blower blows air into the building. The airtightness of the pressurized building is then deduced by monitoring the tracer gas. Keep in mind that the airtightness, or uncontrolled air exchange, is a strong function of wind, temperature, and leakage area. In other words, the air change can vary significantly with changes in outdoor (-indoor) conditions.

One can see that direct measurement is laborious and costly. Mainly for this reason, it is only utilized under special circumstances. It is more practical to estimate the air exchange indirectly. The **Crack Method** is an accepted approach to estimate air exchange. The idea behind this is that if the openings or cracks of a building are known, the air exchange rate can then be expressed as a function of the outside-to-inside pressure difference. The volume flow rate through an opening,

$$\forall' = A\, c\, \Delta P^n, \tag{6.6}$$

where A is the area of the opening, c is the flow coefficient, ΔP is the pressure difference between the outdoor pressure, P_o, and the indoor pressure, P_i, $\Delta P = P_o - P_i$, and n is an empirical exponent with a value that falls between 0.4 and 1. The total flow rate is obtained by summing all k openings as

$$\dot{\forall} = \sum_k A_k c_k \Delta P_k^{n_k}, \tag{6.7}$$

for all terms with ΔP greater than 0, which signifies infiltration. The negative pressure terms represent exfiltration, and under steady-state conditions, where there is no undesirable over-pressurization or under-pressurization, they balance out the infiltration.

Data obtained from particular buildings can be extended to other similar buildings. This will probably suffice for all general purposes. When a more precise estimate is needed, one could resort to modeling. The modeling approach is more precise and it can readily reveal the underlying details, allowing proper measures to be implemented to reduce undesirable air exchange. However, a fair amount of effort is needed to ensure the proper working of the model. To ensure accuracy, verification with real data is necessary.

The outdoor–indoor pressure difference results from three effects [Kuehn et al., 1998, Kreider et al., 2002, McQuiston et al., 2005],

$$\Delta P = \Delta P_w + \Delta P_s + \Delta P_p, \tag{6.8}$$

where ΔP_w is the pressure difference due to wind, ΔP_s is the pressure difference due to stack effects, and ΔP_p is the pressure difference due to building pressurization. Let us examine each effect separately and then combine them to depict a typical real situation.

6.4.1 Pressure Difference due to Wind

Ideally, the air velocity goes to zero as the air approaches the surface of a building. At this (stagnation) point, all the kinetic energy is converted into pressure,[2] boosting the pressure to its maximum value, i.e.,

$$\Delta P_{w,m} = \frac{1}{2}\rho\, V^2. \tag{6.9}$$

We see that this equation is one of the common expressions of the Bernoulli equation, where the potential or elevation effect is negligible. Introducing an empirical parameter, C_p, the pressure coefficient, to account for non-ideal real situations gives

$$\Delta P_w = C_P\, \Delta P_{w,m} = \frac{1}{2}C_P\rho\, V^2. \tag{6.10}$$

Figure 6.2 depicts the typical wind-induced pressure coefficient distribution around a low-rise building.

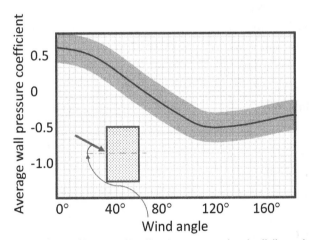

Figure 6.2. Pressure coefficient distribution around a building due to wind for a low-rise building (created by D. Ting). Note that the pressure coefficient is zero when the wind is at approximately 60 degrees, where the building is most streamlined with respect to the wind.

---

[2]Pressure is flow energy, as discussed in Chapter 2.

6.4.2 Pressure Difference due to Stack Effect

The mass of the air column above a particular elevation gives rise to the local air pressure. Thereupon, the atmospheric air pressure decreases with elevation. This decrease is small over the height of typical buildings. More importantly, it is significantly more than the change in indoor pressure with height as a consequence of rising hot air, which pushes from inside out at the upper level, and the corresponding 'vacuum' created at the lower level. In other words, the change in indoor air pressure is primarily caused by the decrease in air density with height, as warmer air rises, leaving the cooler air at the bottom. Ideally, the local hydrostatic outdoor–indoor pressure difference due to air-density differences is

$$\Delta P_s = \Delta H \, g \, (\rho_o - \rho_i),\qquad(6.11)$$

where ΔH is the height from the neutral pressure level, g is the gravity, ρ_o is the density of the outdoor air, and ρ_i is the density of the indoor air. With temperature stratification, cool to warm from the floor to the ceiling, the outdoor–indoor pressure difference varies from positive to negative with height, see Fig. 6.3. By this very nature, outdoor air leaks into the building through the bottom part, and indoor air leaks out of the building through the upper portion of the building. This is more so in the wintertime, when the indoor-outdoor air temperature is large. The stack effect is less significant in the summer, where the cooled air tends to sink to the lower level inside the building. The corresponding larger indoor–outdoor temperature difference, compared to the smaller indoor–outdoor temperature difference at the upper level, may even result in the exfiltration of the cooler air via the lower portion of the building.

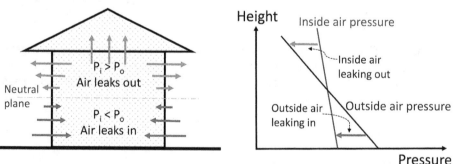

Figure 6.3. Ideal outdoor–indoor pressure difference due to the stack effect (created by D. Ting). The pressure decreases with elevation. This decrease is significantly lessened indoor due to the rising hot air, which increases the pressure at the upper level and decreases the pressure at the lower level.

Assuming an ideal gas,

$$\Delta P_s = \Delta H \, g \, [P_o/(R_o T_o) - P_i/(R_i T_i)], \tag{6.12}$$

where R's are the gas constants, and the temperatures are in absolute scale. Since $R_o \approx R_i \approx R_a$ and $P_o \approx P_i \approx P$, therefore,

$$\Delta P_s = (\Delta H \, g \, P/R_a)(1/T_o - 1/T_i), \tag{6.13}$$

or

$$\Delta P_s = \Delta H \, g \, \rho \, (T_o - T_i)/T_o. \tag{6.14}$$

Introducing an empirical parameter, C_s (the thermal draft coefficient), to account for the non-ideal situation (vertical separations such as floors), gives the actual pressure difference,

$$\Delta P_s = C_s \, \Delta H \, g \, \rho \, (T_o - T_i)/T_o. \tag{6.15}$$

Note that C_s is less than or equal to 1.

EXAMPLE 6.2. OUTDOOR–INDOOR PRESSURE DIFFERENCE

Given: A 200-m tall building with indoor air at 20°C, where the outdoor is at 0°C and 1 atm.

Find: The outdoor–indoor pressure difference at 50 m and 125 m.

Solution:

$$\Delta P_s = C_s \, \Delta H \, g \, \rho \, (T_o - T_i)/T_o,$$

where T is in an absolute scale.

Assume $C_s = 1$, i.e., air is completely free to move (vertically) inside the building. The neutral height is 100 m.

$$T_o = 0°C = 273.15 \, K, \, \rho = 1.29 \, kg/m^3$$
$$T_i = 20°C = 293.15 \, K, \, \rho = 1.20 \, kg/m^3$$

Approximately, $\rho \approx 1.25 \, kg/m^3$.

(Continued)

At 50 m,

$$\Delta P_s = C_s \Delta H g \rho (T_o - T_i)/T_o$$
$$= (1)(50 - 100)(9.81)(1.25)(273.15 - 293.15)/273.15$$
$$= 45 \text{ Pa}$$

At 125 m,

$$\Delta P_s = C_s \Delta H g \rho (T_o - T_i)/T_o$$
$$= (1)(125 - 100)(9.81)(1.25)(273.15 - 293.15)/273.15$$
$$= -22 \text{ Pa}$$

In short, below the neutral height of 100 m, the outside air is at a higher pressure, and thus, it leaks into the building. The opposite is true above 100 m.

Special care must be exercised for high rises with underground garages. The stack effect could cause outdoor air to be sucked into the garage, even through the operating garage fans at times. Along with the outdoor air, the exhaust from the vehicles can be pulled up through the elevator shaft, causing poisonous carbon monoxide to leak through the apartments on the top levels, especially the penthouse.

6.4.3 Pressure Difference due to Building Pressurization

Negative pressurization of buildings is most commonly encountered when powerful exhaust fans are turned on. Within a building, proper air balancing among the different conditioning zones is also essential. In general, the air distribution of a new building is balanced before occupation. Significant changes, especially the adjustment of the temperatures of particular occupied spaces, can substantially upset the air balance. The resulting effect is zone-to-zone pressurization and depressurization. Depressurization that leads to undesirable air flow, such as outdoor air being sucked through the flue pipes can be deadly. Every now and then there are campers who are killed by the silent killer, carbon monoxide, because the attic or exhaust fan of the cabin that they are staying is turned on when the furnace is in operation, and the windows are tightly closed.

6.4.4 The Combined Effect of ΔP (=ΔP_w + ΔP_s + ΔP_p) on Air Leakage

The overall effect of pressure differences on air exchange can be estimated from the summation of the wind-induced, stack-induced, and pressurization-induced

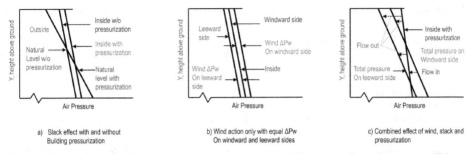

a) Slack effect with and without Building pressurization

b) Wind action only with equal ΔPw On windward and leeward sides

c) Combined effect of wind, stack and pressurization

Figure 6.4. A typical outdoor–indoor pressure difference for winter conditions (created by S. Akhand).

outdoor-to-indoor pressure differences. Figure 6.4 depicts the typical pressure distribution of a building for winter conditions. The key is that the three effects could be complementary or countering. When they are complementary, the leakage is maximized, whereas when they are countering one another, the effects tend to cancel, minimizing the air exchange.

Lawrence Berkeley Laboratory used pressurization tests with an indoor–outdoor pressure difference of 4 Pa [Kuehn et al., 1998]. They developed the following empirical expression:

$$\forall' = A_{\text{leak}} \left(a_s\, \Delta T + a_w V_w^2 \right)^{0.5}, \tag{6.16}$$

where A_{leak} is the effective leak area, a_s is the stack coefficient, and a_w is the wind coefficient. Following Kuehn et al. [1998] and Kreider et al. [2002], first estimates of stack and wind coefficients are tabulated in Tables 6.2 and 6.3, respectively. Table 6.4 lists some estimates of effective leakage area, following Kuehn et al. [1998] and Kreider et al. [2002].

Table 6.2. Stack coefficient for outdoor–indoor pressure difference due to stack effect.

	Number of stories		
	1	2	3
Stack coefficient $[(\text{L/s})^2/(\text{cm}^4 \cdot {}^\circ\text{C})]$	0.000145	0.000290	0.000435

Table 6.3. Wind coefficient for outdoor–indoor pressure difference due to wind effect.

| Shielding class | Wind coefficient in $(L/s)^2/[cm^4 \cdot (m/s)]$ | | |
| | Number of stories | | |
	1	2	3
1. No shielding	0.000319	0.000420	0.000494
2. Light shielding; a few trees or small obstructions	0.000246	0.000325	0.000382
3. Moderate shielding; some obstructions within two house heights, thick hedge, solid fence, or one nearby house	0.000174	0.000231	0.000271
4. Heavy shielding; typical suburban shielding	0.000104	0.000137	0.000161
5. Very heavy shielding; typical downtown shielding	0.000032	0.000042	0.000049

Table 6.4. Sample effective leakage areas.

Building component	Effective leakage area (cm^2/m) or (cm^2/m^2)
Door, single door without weatherstripping (cm^2/m^2)	11
Door, single door with weatherstripping (cm^2/m^2)	8
Door frame-wall, uncaulked masonry wall (cm^2/m^2)	5
Door frame-wall, caulked masonry wall (cm^2/m^2)	3
Wall-ceiling, untaped (cm^2/m)	1
Wall-ceiling, taped (cm^2/m)	0
Wall-sill foundation, uncaulked (cm^2/m)	4
Wall-sill foundation, caulked (cm^2/m)	1

(Continued)

Building component	Effective leakage area (cm^2/m) or (cm^2/m^2)
Window, single-hung without weatherstripping (cm^2/m^2)	4
Window, single-hung with weatherstripping (cm^2/m^2)	2
Window frame-wall, uncaulked masonry wall (cm^2/m^2)	5
Window frame-wall, caulked masonry wall (cm^2/m^2)	1

EXAMPLE 6.3. BUILDING AIR LEAKAGE

Given: A simple rectangular building is 20 m long by 10 m deep by 2.5 m high. It has a flat roof and four 1 m by 1 m windows and two 2.5 m by 1 m doors.

Find: **a.** The effective leakage area.

b. The air exchange rate when $T_i = 20°C$ & $T_o = -20°C$ on a typical winter day.

Solution:

Table 6.5. Building component leakage estimates.

Component	Area (m^2) or perimeter (m)	Leakage area per area or perimeter	Leakage area (cm^2)
Door × 2	2.5 m^2 × 2	12 cm^2/m^2	60
Door frame-wall × 2	6 m × 2	5 cm^2/m^2	30
Wall-ceiling	60 m	1 cm^2/m	60
Wall-sill	60 m	4 cm^2/m	240
Window × 4	1 m^2 × 4	4 cm^2/m^2	16
Window frame-wall × 4	4 m × 4	5 cm^2/m^2	80
Total			486

(*Continued*)

$$\forall' = A_{leak} \left(a_s \, \Delta T + a_w V_w^2\right)^{0.5},$$

From Table 6.1, for a single-story building, $a_s = 0.000145$ $(L/s)^2/(cm^4 \cdot °C)$.

Assuming moderate shielding, from Table 6.2, $a_w = 0.000174$ $(L/s)^2/[cm^4 \cdot (m/s)]$.

$$\begin{aligned}
\forall' &= 486 \, cm^2 \{0.000145 \, (L/s)^2 \, / \, (cm^4 \cdot °C) \, 40°C \\
&\quad + 0.000174 \, (L/s)^2 \, / \, [cm^4 \cdot (m/s)] \, 6.7 \, m/s\}^{0.5} \\
&= 41 \, L/s
\end{aligned}$$

In terms of air exchange rate,

$$\begin{aligned}
A_{CH} &= \forall' / \forall_{building} \\
&= 41 \, L/s / \, (20 \, m \times 10 \, m \times 2.5 \, m) \\
&= 0.041 \, m^3/s \times 3600 \, s/h/500 \, m^3 \\
&= 0.3
\end{aligned}$$

PROBLEMS

Problem 6.1

A wall consists of 1.8 cm of gypsum plaster, 10 cm of glass wool insulation, and 10 cm of facebrick.

(A) What is the U-value of the wall for no wind, summer design wind, and winter design wind?

(B) What is the inside (indoor wall) surface temperature if the indoor air is at 20°C and the outdoor air is at −10°C?

Problem 6.2

The base of a two-story house of 5 m height is subjected to a typical winter wind of 6.7 m/s, resulting in a pressure coefficient, C_P, of 0.5. If the indoor is at 20°C and the outdoor is at −10°C, what is the corresponding outdoor–indoor pressure difference?

Problem 6.3

Draw a line showing the typical indoor air pressure variation with respect to height. The vertical axis is height and the horizontal axis is pressure. Also show the typical outdoor air pressure variation on the same figure. Show and explain infiltration and exfiltration.

Problem 6.4

Apply caulking, weatherstripping, and taping to the house in Example 6.3. What is the resulting A_{CH}? How much heating energy can be saved in a day, assuming the weather conditions remain the same?

Problem 6.5

Assume the building in Example 6.2 has evenly distributed cracks with an effective leakage area of 500 cm² per floor. What is the corresponding infiltration due to the stack effect?

REFERENCES

American Society of Heating, Refrigeration and Air Conditioning Engineers (ASHRAE), *1997 ASHRAE Handbook: Fundamentals,* SI Edition, Atlanta, 1997.

American Society of Heating, Refrigeration and Air Conditioning Engineers (ASHRAE), *2017 ASHRAE Handbook: Fundamentals,* SI Edition, Atlanta, 2017.

R.J. Cole, N.S. Sturrock, "The convective heat exchange at the external surface of buildings," Building and Environment, 12: 207–214, 1977.

E. de Oliveira Fernandes, "Indoor air quality: insights for designing energy efficient buildings," International Journal of Solar Energy, 15: 37–45, 1994.

D.T. Harrje, R.A. Grot, D.T. Grimsrud, "Air infiltration site measurement techniques," 2nd AIC Conference, Building Design for Minim Air Infiltration, Royal Institute of Technology, Stockholm, Sweden, September 21–23, 1981.

J.F. Kreider, P.S. Curtiss, A. Rabl, *Heating and Cooling of Buildings: Design for Efficiency,* 2nd ed., McGraw-Hill, New York, 2002.

T.H. Kuehn, J.W. Ramsey, J.L. Threlkeld, *Thermal Environmental Engineering,* 3rd ed., Prentice-Hall, Upper Saddle River, 1998.

F.C. McQuiston, J.D. Parker, J.D. Spitler, *Heating, Ventilation, and Air Conditioning: Analysis and Design,* 6th ed., Wiley, Hoboken, 2005.

S. Sharples, P.S. Charlesworth, "Full-scale measurements of wind-induced convective heat transfer from a roof-mounted flate plate solar collector," Solar Energy, 62(2): 69–77, 1998.

M.H. Sherman, D.T. Grimsrud, P.E. Condon, B.V. Smith, "Air infiltration measurement techniques," The First International Energy Agency Symposium of the Air Infiltration Centre, Windsor, England, October 6–8, 1980.

F.L. Test, R.C. Lessmann, A. Johary, "Heat transfer during wind flow over rectangular bodies in the natural environment," Journal of Heat Transfer, 103: 261–267, 1981.

CHAPTER 7

Solar Radiation

*"We are like tenant farmers chopping down the fence around our house
for fuel when we should be using Nature's inexhaustible sources of
energy – sun, wind and tide... I'd put my money on the sun and solar energy.
What a source of power!"*

–Thomas Edison

CHAPTER OBJECTIVES

- Appreciate the Sun, without which there is no life on Earth.
- Comprehend the specifics of the tilted Earth rotating and revolving around the Sun.
- Locate places on Earth with reference to the Sun.
- Differentiate the local time from the solar time.
- Recognize Sun's zenith, altitude, and azimuth angles.
- Understand surface azimuth, surface tilt, and surface-solar azimuth.
- Learn to properly utilize overhangs, setbacks, and shadings.
- Fathom solar irradiance and its direct, diffused, and reflected components.
- Estimate solar heat gain through fenestration.

Nomenclature

A Area; A_{ground} is the area of the ground

A_{DS} The apparent direct normal solar flux at the outer edge of the earth's atmosphere

a Window height

a_{coef} Absorption coefficient

a_λ Fraction available after absorption

B   A coefficient associated with the equation of time, which is related to the day of the year, and hence, sun's declination

B_{ext} The apparent atmospheric extinction coefficient

b   Setback depth

C   The clearness number (C = 1 on a clear day)

c   Window width

CT   Clock time

d   Sun's declination, positive when the sun is north of the equator

DSA   Double-strength sheet glass

DT   Daylight savings time correction

E   Equation of time

F_{gA} The shape (view) factor from the ground to the area A

F_s Sunlit fraction

f   Extent of the overhang

g   The horizontal stretch of the overhang beyond the window width

h   Hour angle, negative before solar noon and positive after solar noon, or (convective) heat transfer coefficient; h_i is the interior heat transfer coefficient; h_o is the outside heat transfer coefficient

HVAC   Heating, ventilation, and air conditioning

I   Solar irradiance; $I_{diffuse}$ is the diffused irradiance, $I_{diffuse,\ horizontal}$ is the diffused irradiance on a horizontal surface, $I_{diffuse,\ vertical}$ is the diffused irradiance on a vertical surface, I_{direct} is the direct irradiance, $I_{direct,\ horizontal}$ is the direct irradiance on a horizontal surface, I_f is the final solar irradiance reaching the ground or building, $I_{horizontal,total}$ is the total irradiance on a horizontal surface, $I_{N,o} = 1367\ W/m^2$ (432 Btu/hr·ft^2) is the solar constant, I_o is extraterrestrial irradiance, $I_{reflect}$ is the reflected irradiance, I_{tot} is the total irradiance, I_λ is the solar irradiance corresponding to wavelength λ, $I_{\lambda,f}$ is the solar irradiance corresponding to wavelength λ, which reaches the ground, $I_{\lambda,o}$ is the original solar irradiance corresponding to wavelength λ

K   Extinction coefficient; K_λ is monochromatic extinction coefficient corresponding to wavelength λ

L   Thickness

l   Latitude, positive if it is north of the equator

(Continued)

L_{loc} Local latitude, degrees west

L_{std} Standard meridian for the local time zone, degree west

LST   Local solar time

n   The day of the year

N_i Fraction of the absorbed radiation that becomes a heat gain on the inside, $N_i = h_i/(h_i + h_o)$

n_{ref} Index of refraction

$q_i{}'$ Interior heat-gain flux

r   Reflected fraction

SC   Shading coefficient

SHGC   Solar heat-gain coefficient factor; $SHGC_{diffuse}$ is the solar heat-gain coefficient of the diffused radiation; $SHGC_{direct}$ is the solar heat-gain coefficient of the direct radiation; $SHGC_{horizontal}$ is the solar heat-gain coefficient for a horizontal surface; $SHGC_{reflected,corrected}$ is the solar heat-gain coefficient of the reflected radiation after correction for deviation of ground reflectivity from 0.2; $SHGC_{table}$ is the tabulated solar heat-gain coefficient at standard conditions

T   Temperature; $T_{glass,i}$ is the interior glass temperature, T_i is the indoor (interior) temperature; T_o is the outdoor (outside) temperature

U   Heat transfer coefficient, or, U factor

x   Shadow width

y   Distance in the vertical direction

z   Thickness

Greek and Other Symbols

α Absorptivity; α_λ is the total monochromatic absorptivity at wavelength λ

β Sun's altitude angle; β_{noon} is sun's altitude at solar noon

γ Surface-solar azimuth, the angle between the horizontal projection of the solar rays and the horizontal projection of the surface normal

δ Profile angle, projected altitude angle

θ Zenith angle; θ_H is the zenith angle for a horizontal surface

λ Wavelength

ρ Reflectivity; ρ_{ground} is the reflectivity of the ground, ρ_λ is the total monochromatic reflectivity at wavelength λ

Σ Surface tilt, the angle between the surface normal and vertical

(Continued)

τ Transmissivity; τ_{avg} is the average transmissivity; $\tau_{diffuse}$ is the transmissivity of the diffused radiation; τ_{direct} is the transmissivity of the direct radiation; $\tau_{reflected}$ is the transmissivity of the reflected radiation; τ_λ is the total monochromatic transmissivity at wavelength λ

φ Azimuth angle, the sun's location with respect to south, negative when it is east of south, and positive when it is west of south

Ψ Surface azimuth angle, positive west of south and negative east of south

7.1 OUR GOOD SUN

Without the Sun, Earth would be but a frozen, lifeless planet. It is a humbling experience to accept that we revolve around the Sun, i.e., the Sun does not revolve around us. Furthermore, we feed on the Sun, and never ever give anything back. The term "photosynthesis" alone should make us stand in awe of the accuracy of flaked-cereal genius, Dr. John Harvey Kellogg's statement, "Food is simply sunlight in cold storage."

In the absence of solar radiation, circadian rhythms (24-hour cycle in the physiological processes of living things such as plants and animals) would be disrupted to the extent that could completely wipe out life on earth. From a heating, ventilation, and air conditioning (HVAC) point of view, heat and light from the Sun are also indispensable elements. In the winter time, both heat and light are precious. Effectively embracing heat and light into the indoor environment can lead to significant savings on humanmade, energy-intensive heating and lighting energy usage, and also simultaneously uplifting the spirit from the winter blues. In the summer, heat is generally an unwanted guest inside the building, but light is still welcome. However, the tide on rejecting the summer solar heat is turning. It is becoming clear that we should always welcome the sun by properly harnessing its energy to reduce the cooling energy demand and save our environment. Solar cooling, whenever feasible, can make a big stride toward both fronts. Other than the high-capital cost, bulkiness is also a stumbling block for solar cooling systems. Solar photovoltaic technology, on the other hand, is a slick means to rein in the solar energy into electricity. This high-quality energy can be employed to power all kinds of engineering systems for sustaining high-quality living. An emerging technology is creating building facades, which are promising engineering innovations for leapfrogging the concurrent capitalization of sunlight and solar energy.

Talking about waste heat, every system that is powered by electricity rejects an amount of heat equal to the system's power rating. This is not too bad during the cold season as the generated heat contributes to the heating, lowering the needed heating supplied by the engine to power the electric heating. Heating powered by high-quality energy, electricity, is never a good idea from the perspective of the second law of thermodynamics. In plain English, we should always avoid using high-quality energy (electricity) to perform the very low-quality task of heating. Along this line, converting from conventional incandescent lights to fluorescent lights can lead to significant savings in terms of supplying the same amount of light with much less energy. This replacement enables furnishing the same amount of brightness while generating less waste heat. Other than the cold season, the waste heat needs to be removed from the occupied space. Today, many fluorescent lights are being replaced by even better engineering inventions, i.e., light-emitting diode (LED) lights. What remains to be improved are high-heat-dissipating spotlights, where the transition into LED ones is somewhat sluggish, for whatever reasons.

In the calculation of heating and cooling loads, we are concerned with (a) peak heating load and (b) peak cooling load. The peak heating load sets the capacity of the heating system. As mentioned earlier, the peak heating demand occurs during cold winter nights, where the contribution of solar radiation can be neglected. Solar radiation is important when it comes to annual energy consumption, nevertheless. The peak cooling load, on the other hand, specifies the cooling system capacity. Since the maximum cooling demand takes place during hot sunny days, the solar contribution is crucial. Other than these peak demands, the day-to-day usage is also important, as they add up to the total annual energy consumption.

7.2 THE SUN–EARTH RELATION

The earth has a diameter of approximately 12.7×10^3 km. It rotates about its polar axis every 24 hours. The earth also revolves around the sun in a roughly elliptic orbit, see Fig. 7.1, completing one revolution per year ($\approx$365.25 days). The mean earth–sun distance is roughly 1.5×10^8 km, and this distance is shortest around January 1, resulting in 7% more radiation than in July [Kuehn et al., 1998; McQuiston, et al., 2005]. Northern hemisphere habitants should thank the good heaven, for otherwise their winter would be much colder! The earth is farthest away from the sun around July 1, leading to a reduction in solar radiation by 3.3%. Once again, the inhabitants of the Northern hemisphere lucked out, for otherwise the summer would be much more unenjoyable in the Northern hemisphere! No wonder Isaac

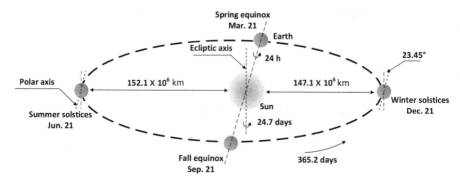

Figure 7.1. The Sun–Earth Relation (created by X. Wang). We have equal length of day and night during the equinoxes.

Newton exclaimed, "When I look at the solar system, I see the earth at the right distance from the sun to receive the proper amounts of heat and light. This did not happen by chance."

The axis of rotation (polar axis) of the earth is tilted 23.5° with respect to its orbit around the sun. This gives rise to the summer and winter solstices. Because of this tilt, the earth's surface can be divided into the following:

1) Torrid Zone. The torrid zone encompasses ±23.5° from the equator (the tropic). Within the torrid zone the sun is at the zenith (vertically overhead) at least once yearly.

2) Temperate Zones. The two temperate zones cover latitudes from 23.5 to 66.5°, with one north of the equator and the other south of it. In the temperate zones, the sun appears above the horizon each day but never at the zenith.

3) Frigid Zones. The region from the north pole to 66.5° north is the northern frigid zone, and it is called the arctic circle. Its counterpart in the south is called the antarctic circle. The sun is below the horizon (and above) for at least one full day yearly.

7.3 MAPPING THE EARTH WITH RESPECT TO THE SUN

The easiest way to describe a location on earth is to provide its latitude and longitude (see Fig. 7.2). The latitude and longitude are directly related to the orientation of the earth with respect to the sun.

7.3.1 Latitude, Hour Angle, and Sun's Declination

To specify the direction of the sun's rays at a location on earth, say point P, we need the latitude, l, the hour angle, h, and the sun's declination, d. These angles

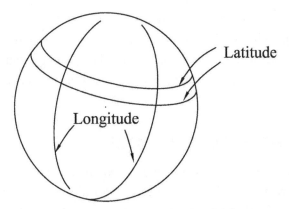

Figure 7.2.   Latitude and longitude (created by N. Cao).

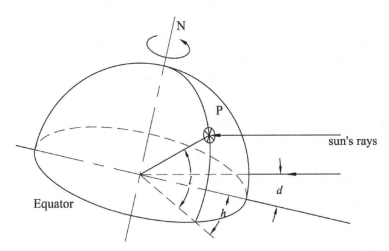

Figure 7.3.   Latitude, hour angle, and sun's declination (created by N. Cao). The hour angle is with respect to the latitude, which is at solar noon.

are depicted in Fig. 7.3. The sign for the latitude, l, is positive when it is north of the equator, and it is negative when it is south of the equator. The hour angle, h, expresses the time of day with respect to solar, not clock, noon. With 360 degrees making a full circle around the earth, dividing it by the 24-hour day leads to 15° per hour, i.e., 15°-hour angle. The sun's declination, d, is the angular distance of the sun's rays north or south of the equator. It is taken as positive when the sun's rays are north of the equator, and negative when they are south of the equator.

The sun's declination can be approximated as

$$d = 23.45 \sin\left[360\left(284 + n\right)/365\right]. \qquad (7.1)$$

where n is the day of the year. The sun's declination angle is equal to 0 during the equinoxes (equal night); see Fig. 7.1. The autumnal equinox for the northern hemisphere falls on September 21, and thus, the official beginning of fall. Vernal equinox takes place on March 23, and the northern habitants welcome it as the official beginning of spring.

7.4 CLOCK TIME VERSUS SOLAR TIME

What time is it? That depends not only on where you are, but also what time you are interested in. The Universal time, also called the Greenwich civil time is the time along the Greenwich meridian (zero longitude). With this reference, local time is retarded by 1/15 h for each degree of longitude west of Greenwich, i.e., 24 h/360°. For ensuring all residents in a particular region, country, state, or province are "on the same page" as far as the local time is concerned, the local time may span beyond the 15°-longitude span. The **local solar time** is also referred to as the **clock time** or the **standard time**.

Solar time is based on the apparent motion of the sun, as seen from a point on the surface of the earth. Succinctly, solar noon is the time when the sun reaches the highest point in the sky, even though this highest sun may be quite a bit shy of vertically overhead during the winter time. When it comes to death, every second counts, and thus, many old worlds relied on the solar time measured with a stick erected vertically from the ground for execution.

The four time zones in North America are the following:

- Eastern Standard Time. This is with reference to 75° west of Greenwich meridian.
- Central Standard Time. Central standard time is employed for regions west of the eastern standard time zone, ideally 90° west of Greenwich meridian.
- Mountain Standard Time. The mountainous terrain is situated around 105° west of Greenwich meridian.
- Pacific Standard Time. The west coast faces the Pacific Ocean, and the region correspond roughly to 120° west of Greenwich meridian.

Currently, most parts of western Europe and North America advance the clock by 1 hour during the summer half of the year. This advanced time is called the daylight saving time, but what does it save? The idea is to better match human activities with the availability of daylight, and hence, reduce overall energy consumption. It is interesting to note that countries such as Sweden celebrates the longest day, whereas others such as Iran commemorate the longest night. More interesting, changes of clock time tend to disrupt human operation. More accidents occur the day after the "spring forward" in the spring, when people lose an hour of sleep. Thus, it is smart for places such as Saskatchewan, Canada, to not spring forward or fall back with the tide, i.e., change of the season.

The **equation of time**, which denotes the difference between Local Solar Time (LST) and Local Civil Time (LCT; also called clock time or standard time), can be expressed as

$$E = 0.165 \sin 2B - 0.126 \cos B - 0.025 \sin B \text{ hours},\qquad (7.2)$$

where $B = 360\,(n - 81)/364$. The Local Solar Time (LST),

$$LST = CT + (1/15)\,(L_{std} - L_{loc}) + E - DT,\qquad (7.3)$$

where LST = Local Solar Time, hr (24 hr format);
  CT = Clock Time, hr (24 hr format);
  L_{std} = standard meridian for the local time zone, degree west;
  L_{loc} = longitude of actual location, degrees west;
  E   = Equation of Time, hr;
  DT = Daylight Savings Time Correction, hr.

Accordingly, the hour angle can be determined as follows:

$$h = 15\,(LST - 12) \text{ degrees.}\qquad (7.4)$$

EXAMPLE 7.1. SOLAR TIME FROM CLOCK TIME

Given: Windsor, Ontario, Canada, at 1:00 p.m. Eastern Daylight Savings Time on June 12.
 Find: Solar time.

Solution:

From Eq. 7.2, the equation of time,

$$E = 0.165 \sin 2B - 0.126 \cos B - 0.025 \sin B \text{ hours},$$

where

$$B = 360\,(n - 81)/364 = 80°$$

(*Continued*)

Therefore,

$$E = 0.01 \approx 0$$

From Eq. 7.3, the Local Solar Time,

$$LST = CT + (1/15)(L_{std} - L_{loc}) + E - DT$$

For Windsor, $L_{std} = 75°$ (Eastern Standard Time) and $L_{loc} = 83.0°$. Therefore,

$$LST = 13 + (1/15)(75 - 83) + 0 - 1 = 11.46 \text{ hour or } 11{:}28 \text{ a.m.}$$

7.5 SURFACE-SOLAR ANGLES

The solar heat gain via the external building envelope elements, roofs, walls, and, most significant of all, windows, is a direct function of the angle of the solar rays hitting the surface. This is especially so when we deal with the amount of solar radiation entering the building through fenestration. As such, we would like to design the various building structures involved to passively and actively control the optimal radiation.

7.5.1 For a Horizontal Surface

Figure 7.4 depicts the sun's zenith, altitude, and azimuth angles. The **zenith angle**, θ_H, is the angle between the sun's rays and local vertical. In a nutshell, the zenith angle indicates how close the sun to the upright direction. The angle in the vertical plane between the sun's rays and the projection of the sun's rays on the horizontal plane is the **altitude angle**, β. In other words, the altitude angle signifies how high the sun is. Note that the sum of the zenith angle and the altitude angle is 90°, i.e.,

$$\theta_H + \beta = \pi/2. \tag{7.5}$$

The units for the angles in Eq. 7.5 are radians. The **azimuth angle**, φ, denotes where the sun is with respect to south. It is the angle in the horizontal plane measured from south to the horizontal projection of the sun's rays. It is negative when it is east of south and positive when it is west of south.

The following are some useful relationships between the various angles we have discussed:

$$\cos \theta_H = \cos l \cos h \cos d + \sin l \sin d, \tag{7.6}$$

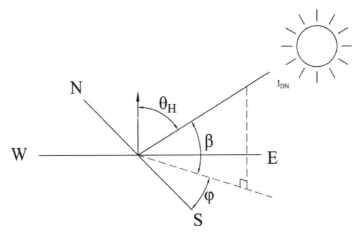

Figure 7.4. Sun's zenith, altitude, and azimuth angles (created by N. Cao).

$$\sin \beta = \cos l \cos h \cos d + \sin l \sin d, \tag{7.7}$$

and

$$\cos \varphi = (1/\cos \beta)(\cos d \sin l \cos h - \sin d \cos l). \tag{7.8}$$

This last expression will not provide the correct sign convention, and hence, the correct sign must be introduced manually. The sign convention is as follows:

Latitude, l	positive for latitudes north of the equator;
Sun's declination, d	positive when the sun's rays are north of the equator;
Hour angle, h	negative before solar noon, and positive after solar noon;
Azimuth angle, φ	negative when it is east of south, and positive when it is west of south.

For rapid determination of the daily maximum altitude of the sun for a given location, we note that this corresponds to solar noon, and thus,

$$\beta_{\text{noon}} = 90° - |l - d|. \tag{7.9}$$

The sun's angle of incidence, θ, is the angle between the solar rays and the surface normal.

EXAMPLE 7.2. SUN'S ALTITUDE AND AZIMUTH

Given: A place at 40° North on August 21.
 Find: **(a)** Sun's altitude and azimuth angles at 7:30 a.m. solar time.

(b) Solar time and azimuth angle for sunrise.

Solution:

(a) From Eq. 7.7,

$$\sin \beta = \cos l \cos h \cos d + \sin l \sin d$$

We have $l = 40°$ and LST = 7.5.
From Eq. 7.4, solar hour angle,

$$h = 15\,(\text{LST} - 12) = -67.5°$$

From Eq. 7.1, the sun's declination for August 21,

$$d = 23.45 \sin\,[360\,(284 + n)/365] = 12.3°$$

Therefore,

$$\sin \beta = \cos l \cos h \cos d + \sin l \sin d = 0.423$$

or

$$\beta = 25.0°$$

From Eq. 7.8, the sun's azimuth angle,

$$\cos \varphi = (1/\cos \beta)(\cos d \sin l \cos h - \sin d \cos l) = 0.085$$

or

$$\varphi = -85°, \text{ where the negative implies East of South.}$$

(b) At sunrise, sun's altitude angle, $\beta = 0$, i.e.,

$$\sin \beta = \cos l \cos h \cos d + \sin l \sin d = 0$$

thus,

$$\cos 40° \cos h \cos 12.3° = \sin 40° \sin 12.3°$$

Rearranging,

$$\cos h = 0.183, \text{ and thus, } h = 79.5 = 15\,(LST - 12)$$

Therefore,

$$LST = 17.3 = 5{:}18 \text{ a.m.}$$

The corresponding solar azimuth angle at sunrise,

$$\cos \varphi = (1/\cos \beta)\,(\cos d \sin l \cos h - \sin d \cos l) = 0.085$$
$$= (1/1)\,(\cos 12.3° \sin 40° \cos 79.5° - \sin 12.3° \cos 40°)$$
$$= -0.0487$$

Hence,

$$\varphi = -92.8°$$

where the negative sign signifies that the sun is east of south, confirming the obvious at sunrise.

7.5.2 For Tilted Surfaces

Horizontal surfaces such as flat roofs are rare, most building surfaces are tilted, and vertical surfaces such as walls and windows are presumably most common. Figure 7.5 illustrates the typical surface-solar angles for a tilted surface. The three key angles are **surface azimuth angle**, Ψ, **surface tilt angle**, Σ, and **surface-solar azimuth angle**, γ.

- Surface azimuth, Ψ. This is the angle between south and the horizontal projection of the surface normal. It denotes which direction the surface is facing with respect to south. It is positive for a surface that faces west of south, and negative when the surface is facing east of south.
- Surface tilt, Σ. The surface tilt is the angle between the surface normal and vertical. A vertical surface has a tilt of zero.
- Surface-solar azimuth, γ. The angle between the horizontal projection of the solar rays and the horizontal projection of the surface normal.

We see that

$$\gamma = |\varphi - \Psi|, \tag{7.10}$$

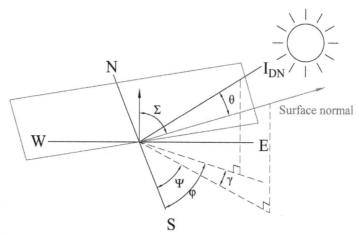

Figure 7.5.   Surface-solar angles for a tilted surface (created by N. Cao).

that is, the surface-solar azimuth is the difference between the solar azimuth (sun's direction with respect to south) and the surface azimuth (surface's direction with respect to south). For the tilted surface, we have

$$\cos \theta = \cos \beta \, \cos \gamma \, \sin \Sigma + \sin \beta \, \cos \Sigma. \tag{7.11}$$

Note that the incidence angle for a horizontal surface is equal to the zenith angle.

7.5.3 Shading and Overhangs

We can see the use of shading, and also overhangs, in some of the oldest human habitats. Over time, their inclusion into buildings has to do with fashion, not just thermal comfort. Properly devised, overhangs and shading can be employed to effectively and passively control the solar heat gains and natural light. The general idea is to exclude direct sun rays in the summer and admit them in the winter. Excluding direct solar irradiance naturally reduces the unwanted heat gain, which requires energy and cooling capacity to remove it to keep the indoor environment cool for thermal comfort. In the winter, maximizing the natural solar heat gain can save significant heating requirements.

Shading generally implies shades that are farther away from the building. Trees and nearby buildings and structures are the most common shading. Overhangs and setbacks are closely associated with the building fenestration. Consider the setback depicted in Fig. 7.6. The shadow depth,

$$y = b \tan \delta, \tag{7.12}$$

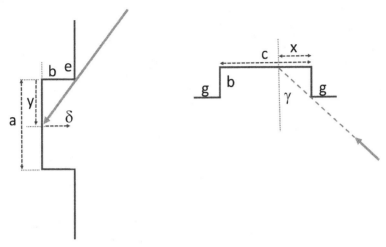

Figure 7.6. Window setback for controlling direct sun rays (created by D. Ting).

where δ is the **profile angle** or **projected altitude angle**, i.e., the angle between horizontal and the projection of the solar rays onto the vertical plane. We can relate the profile angle to sun's altitude angle, β, and the surface-solar azimuth angle, γ; i.e.,

$$\tan \delta = \tan \beta / \cos \gamma. \tag{7.13}$$

It is clear that the shadow width

$$x = b \tan \gamma. \tag{7.14}$$

Therefore, the sunlit fraction

$$F_s = (a - b \tan \delta)(c - b \tan \gamma)/(ac). \tag{7.15}$$

Let us further effectuate the passive control of direct sun rays by integrating an overhang on top of the setback (see Fig. 7.7). For complete shading,

$$x = (b + f) \tan \gamma - g \leq b \tan \gamma, \tag{7.16}$$

or

$$g \geq f \tan \gamma, \tag{7.17}$$

and

$$y = (f + b) \tan \delta - e \geq a. \tag{7.18}$$

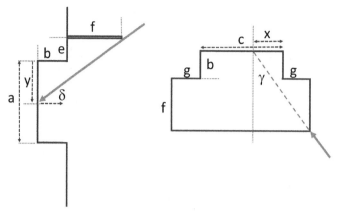

Figure 7.7. Shading of a window using both setback and overhang (created by D. Ting).

In other words, the required extent of the overhang,

$$f \geq (a + e) \cot \delta - b. \tag{7.19}$$

EXAMPLE 7.3. SOLAR SHADING BY AN OVERHANG + SETBACK

Given: A south-facing, 8-ft high window in Crimea, Ukraine (45°N, 33°E), is to be completely shaded at solar noon on summer solstice, and completely sunlit at solar noon on winter solstice.

Find: The length of the overhang from the wall, f, and its distance above the window, e.

Solution:

On summer solstice, sun's declination angle according to Eq. 7.1 is

$$d = 23.45 \sin\left[360\left(284 + n\right)/365\right] = 23.45°.$$

At solar noon, the hour angle, h = 0, and the solar azimuth, $\varphi = 0$. As the window is facing south, the surface azimuth, $\Psi = 0$. The corresponding surface-solar azimuth, $\gamma = |\varphi - \Psi| = 0$. The solar altitude can be deduced from Eq. 7.7,

$$\sin \beta = \cos l \cos h \cos d + \sin l \sin d,$$

which gives $\beta = 68.45°$.

The profile angle can be obtained from Eq. 7.13,

$$\tan \delta = \tan \beta / \cos \gamma,$$

which gives $\delta = 68.45°$.

With $\gamma = 0$, the sunlit fraction can be calculated from Eq. 7.15,

$$F_s = (a - b \tan \delta)(c - b \tan \gamma) / (a c) = (8 - 0)(c - 0) / (8c) = 1,$$

that is, with neither setback nor overhang, the south-facing window is fully sunlit at solar noon.

For the window to be completely shaded, y has to be equal or greater than a. From Eq. 7.18,

$$y = (f + b) \tan \delta - e = a,$$

we can write

$$(f + 0) \tan 68.45° - e = 0.$$

This gives

$$0.395 f = e.$$

Substituting this into $2.532 f = e + 8$, we get

$$2.532 f = 0.395 f + 8,$$

or

$$f = 3.74 \, \text{ft}.$$

Substitute into $2.532 f = e + 8$, we obtain

$$e = 1.48 \, \text{ft}.$$

Now that we know how and when to block or permit the sun's rays through fenestration, let us move on and see how much unwanted and/or wanted solar radiation we are dealing with. In practice, this information may be required to justify the extra cost for installing the overhangs and/or constructing the setbacks. From the knowledge acquired so far, we can position the sun with respect to the window, sunroof, skylight, etc., of concern as a function of the time of the year. What is missing is the amount of solar irradiance as a function of the time of the year and the location and orientation of the window.

7.6 DEPLETION OF EXTRATERRESTRIAL INSOLATION THROUGH THE ATMOSPHERE

To some people, the word "extraterrestrial" subconsciously elicits ET or extraterrestrial beings. As ET refers to beings outside of earth, extraterrestrial insolation signifies the solar irradiance outside of the earth's atmosphere. Our ever-protective atmosphere, which we often take for granted, shields us from many harmful objects. Within context, the ozone layer blocks much of the harmful portion of the sun's rays from harming us. Let us ride on the Magic School Bus and follow the sun's rays into the atmosphere of the earth.

7.6.1 Extraterrestrial Insolation

The solar irradiance outside the earth's atmosphere at normal incidence and at the mean sun–earth distance is called **the solar constant,**

$$I_{N,o} = 1367\,\text{W/m}^2 \left(432\,\text{Btu/hr} \cdot \text{ft}^2\right). \tag{7.20}$$

This is hotter than hot, i.e., it will toast us! Due to the slight eccentricity of the orbit, the actual value of the extraterrestrial irradiance, I_o, varies by ±3.3% over the year. A good fit for I_o is

$$I_o = [1 + 0.033\,\cos(360° \times n/365.25)] \times 1367\,\text{W/m}^2, \tag{7.21}$$

where n = day of year. As suspected from the variation in the earth–sun distance explicated earlier, I_o peaks in the winter of the northern hemisphere. Resultantly, the seasonal hemisphere variations are smaller than they would be if the earth's orbit were circular.

7.6.2 Direct Solar Radiation Reaching the Ground

As a ray from the sun passes through the atmosphere, the beam solar radiation is reduced due to absorption (by ozone in the upper atmosphere and by water vapor nearer to the surface) and scattering. The absorption leads to the heating up of the atmosphere, and the making of much diffuse radiation. We can see from Fig. 7.8 that the radiation depletion increases with (a) radiation intensity and (b) the quantity of material passed through. In other words, the change in I_λ,

$$(I_\lambda - dI_\lambda) - I_\lambda \propto I_\lambda dz. \tag{7.22}$$

This gives

$$-dI_\lambda \propto I_\lambda dy/\sin \beta. \tag{7.23}$$

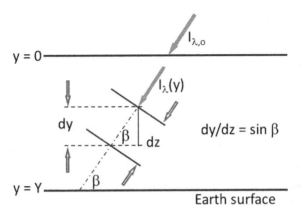

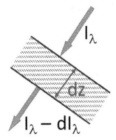

Figure 7.8. Depletion of solar radiation through the atmosphere (created by D. Ting).

Introducing a constant, K_λ (monochromatic extinction coefficient), we have

$$-dI_\lambda = K_\lambda I_\lambda dy/ \sin \beta, \tag{7.24}$$

or

$$dI_\lambda = -K_\lambda I_\lambda (1/ \sin \beta) \, dy. \tag{7.24a}$$

We can integrate this

$$\int (1/I_\lambda) \, d\, I_\lambda = -\int K_\lambda / \sin \beta \, dy = -(1/ \sin \beta) \int K_\lambda dy, \tag{7.25}$$

$$\ln (I_{\lambda, f}/I_{\lambda, o}) = -(1/ \sin \beta) \int K_\lambda \, dy, \tag{7.26}$$

$$I_{\lambda, f} = I_{\lambda, o} \exp \left[-(1/ \sin \beta) \int K_\lambda \, dy \right]. \tag{7.27}$$

This can be rearranged as

$$\frac{I_{\lambda,f}}{I_{\lambda,o}} = \exp\left(-\frac{1}{\sin\beta}\int_0^Y K_\lambda \, dy\right), \tag{7.28}$$

where the right-hand side is the effective monochromatic transmittance.

Over the wavelengths of solar radiation, the shorter wavelengths (higher frequencies) are depleted more readily than the longer ones (see Fig. 7.9). Furthermore, the higher intensity wavelengths are depleted more significantly. Hence, we can apply Eq. 7.24a for the entire solar spectrum,

$$dI = -K I \, dy / \sin\beta. \tag{7.29}$$

Integrating gives

$$I_f = I_o \exp\left(-\frac{1}{\sin\beta}\int_0^Y K \, dy\right) = I_o \exp\left(-\int_0^Y K \, dy / \sin\beta\right) \tag{7.30}$$

We may write this as

$$I_f = C A_{DS} \exp\left[-B_{ext} / \sin\beta\right], \tag{7.31}$$

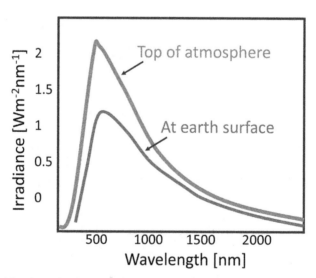

Figure 7.9. Varying depletion of solar radiation spectrum through the atmosphere (created by D. Ting).

Table 7.1. Coefficients for average clear day solar radiation.

Day	A_{DS} (W/m^2)	B_{ext} (–)	C (–)
January 21	1230	0.142	0.058
February 21	1215	0.144	0.060
March 21	1186	0.156	0.071
April 31	1136	0.180	0.097
May 31	1104	0.196	0.121
June 21	1088	0.205	0.134
July 21	1085	0.207	0.136
August 21	1107	0.201	0.122
September 21	1151	0.177	0.092
October 21	1192	0.160	0.073
November 21	1221	0.149	0.063
December 21	1233	0.142	0.057

where C is the clearness number (C = 1 on a clear day), A_{DS} is the apparent direct normal solar flux at the outer edge of the earth's atmosphere, and B_{ext} is the apparent atmospheric extinction coefficient. Sample values of these coefficients, based on Kuehn et al. [1998], which originated from the 1993 ASHRAE Handbook [1993], are tabulated in Table 7.1.

7.7 DIRECT, DIFFUSE, AND REFLECTED RADIATION ON A SURFACE

The total solar radiation striking a surface consists of the direct radiation, the diffuse radiation, and the reflected radiation, i.e.,

$$I_{tot} = I_{direct} + I_{diffuse} + I_{reflect} \qquad (7.32)$$

This is depicted in Fig. 7.10. The direct component is simply the final irradiance reaching the ground multiplied by the cosine of the zenith angle,

$$I_{direct} = I_f \cos \theta. \qquad (7.33)$$

This is generally the largest irradiance on a building surface. For a horizontal surface,

$$I_{direct, horizontal} = I_f \cos \theta_H = I_f \sin \beta, \qquad (7.34)$$

(see Fig. 7.11).

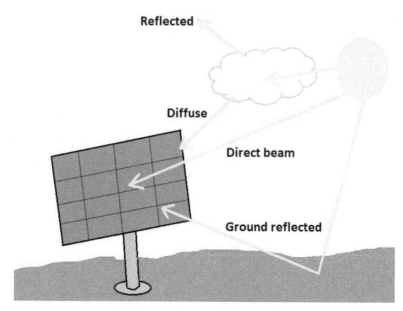

Figure 7.10. Direct, diffuse, and reflected solar radiation on a surface (created by S.K. Mohanakrishnan).

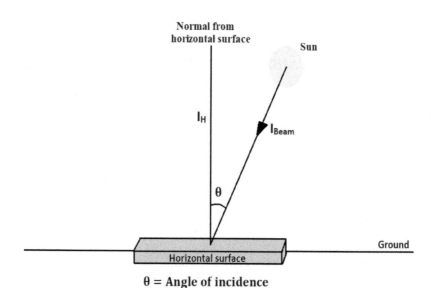

θ = Angle of incidence

Figure 7.11. Direct solar radiation on a horizontal surface (created by S.K. Mohanakrishnan).

The diffuse component, I_{diffuse}, is mostly from the sky. The amount on a vertical surface can be related to that on a horizontal surface as

$$I_{\text{diffuse, vertical}}/I_{\text{diffuse, horizontal}} = 0.45 \text{ for } \cos\theta \leq -0.2, \tag{7.35}$$

$$= 0.55 + 0.437\cos\theta + 0.313\cos^2\theta \text{ for } \cos\theta > -0.2, \tag{7.36}$$

where

$$I_{\text{diffuse, horizontal}} = C\, I_f. \tag{7.37}$$

Typical values of the clearness index can be found in Table 7.1. For non-vertical surfaces,

$$I_{\text{diffuse}} = I_{\text{diffuse, horizontal}}\,(1 + \cos\Sigma)/2. \tag{7.38}$$

The Reflected Solar Radiation

The ground can reflect some amount of solar radiation onto the nearby walls. The reflected amount is dictated by the way the surface of interest and the ground see each other, i.e.,

$$A\, I_{\text{reflected}} = \rho_{\text{ground}}\, I_{\text{horizontal,total}}\, A_{\text{ground}}\, F_{gA}, \tag{7.39}$$

where ρ_{ground} is the reflectivity of the ground, A is area of the surface of concern, A_{ground} is the area of the ground, and F_{gA} is the shape factor from the ground to area A, i.e., fraction of radiation leaving the ground that strikes the surface. The ground reflectivity in the summer time is quite small (~0.1). In the winter time, on the other hand, it can approach unity right after snowing, and stays around unity in a clean environment where the fresh snow remains fresh for a long time. One-side effect of this is "snow burn," where skiers realize that they have a sunburn after enjoying skiing on a clear and sunny day in fresh snow. It is interesting to note that there are studies that suggest that emissions of black soot have been altering the way sunlight reflects off snow, i.e., the albedo effect [Qian et al., 2009; Hadley & Kirchstetter, 2012]. The research along this line indicates that soot could be responsible for a considerable amount of global warming over the past century.

For any two surfaces, say a building surface with area A and the ground with area A_{ground},

$$A\, F_{Ag} = A_{\text{ground}}\, F_{gA}. \tag{7.40}$$

This can be rearranged into

$$F_{Ag} = \left(A_{ground}\, F_{gA}\right)/A. \tag{7.41}$$

Substituting this into Eq. 7.39 and rearranging gives

$$I_{reflected} = \rho_{ground}\, I_{horizontal,\, total} \left(A_{ground}\, F_{gA}\right)/A \tag{7.42}$$

or

$$I_{reflected} = \rho_{ground}\, I_{horizontal,\, total}\, F_{Ag}. \tag{7.43}$$

For a diffusively reflecting, non-vertical surface with a tilt of Σ,

$$F_{Ag} = (1 - \cos \Sigma)/2. \tag{7.44}$$

In other words, for a general surface with a tilt angle of Σ, the radiation from the ground,

$$I_{reflected} = \rho_{ground}\, I_{horizontal,\, total}\, (1 - \cos \Sigma)/2. \tag{7.45}$$

EXAMPLE 7.4. WHEN SNOW RADIATION IS SIGNIFICANT

Given: A building in Windsor, Ontario, Canada, has a skylight that faces 15° east of south and is tilted 60° from the ground.

Find: I_{total} on the skylight at 1:30 p.m. solar time on December 21. Assume that the ground is covered with fresh Christmas snow where $\rho_{ground} = 0.8$.

Solution:

The total radiation, according to Eq. 7.32, is

$$I_{tot} = I_{direct} + I_{diffuse} + I_{reflect}.$$

The direct component on a surface with a tilt of Σ is simply the final irradiance reaching the ground multiplied by the cosine of the incidence angle, as described by Eq. 7.33, i.e.,

$$I_{direct} = I_f \cos \theta.$$

The needed zenith angle can be calculated from Eq. 7.11,

$$\cos \theta = \cos \beta \cos \gamma \sin \Sigma + \sin \beta \cos \Sigma.$$

(*Continued*)

The solar altitude angle, β, can be deduced from Eq. 7.7,

$$\sin \beta = \cos l \cos h \cos d + \sin l \sin d$$

For Windsor, Ontario, Canada, the latitude, l, is 42.3°, $L_{std} = 75°$, $L_{loc} = 83.0°$.
The hour angle, $h = 15$ (LST-12) = 22.5°.
According to Eq. 7.1, the solar declination

$$d = 23.45 \sin [360 (284 + n)/365] = -23.45°.$$

Therefore,

$$\sin \beta = \cos l \cos h \cos d + \sin l \sin d = 0.36$$

which gives $\beta = 21.0°$.
   The surface-solar azimuth,

$$\gamma = |\varphi - \Psi|$$

With the surface faces 15° east of south, $\Psi = -15°$.
The solar azimuth can be computed from Eq. 7.8,

$$\cos \varphi = (1/\cos \beta)(\cos d \sin l \cos h - \sin d \cos l) = 0.9263$$

which gives $\varphi = 22.14°$, i.e., west of south.
   Therefore, $\gamma = 37.14°$.
   With the above, we can compute the zenith angle,

$$\cos \theta = \cos \beta \cos \gamma \sin \Sigma + \sin \beta \cos \Sigma = 0.8237$$

or $\theta = 34.54°$.
   From Eq. 7.31, the final irradiance reaching the ground,

$$I_f = C A_{DS} \exp [-B_{ext}/ \sin \beta].$$

From Table 7.1, we have C = 1, $A_{DS} = 1233$ W/m², and $B_{ext} = 0.142$.
Therefore,

$$I_{direct} = I_f \cos \theta = 683 \text{ W/m}^2.$$

This direct component is typically the largest among the three components.

For a tilted surface, the diffuse component can be estimated from Eq. 7.38,

$$I_{\text{diffuse}} = I_{\text{diffuse, horizontal}} \left(1 + \cos \Sigma\right)/2,$$

where $I_{\text{diffuse, horizontal}} = C\, I_f = 0.057\, (830\ \text{W/m}^2)$. Hence, $I_{\text{diffuse}} = 35\ \text{W/m}^2$. We note that the diffuse component is relatively small, as it typically is. The reflected component, as per Eq. 7.45, is

$$I_{\text{reflected}} = \rho_{\text{ground}}\, I_{\text{horizontal, total}} \left(1 - \cos \Sigma\right)/2.$$

According to Eq. 7.34,

$$I_{\text{direct, horizontal}} = I_f \cos \theta_H = I_f \sin \beta = 297\ \text{W/m}^2.$$

The diffuse component on a horizontal surface can be deduced from Eq. 7.37,

$$I_{\text{diffuse, horizontal}} = C\, I_f = 47\ \text{W/m}^2.$$

With no nearby mirror or large reflective structures reflecting onto the ground,

$$I_{\text{reflected, horizontal}} = 0.$$

Summing the three components gives

$$I_{\text{horizontal, total}} = 344\ \text{W/m}^2.$$

From this, we get

$$I_{\text{reflected}} = 138\ \text{W/m}^2.$$

Therefore,

$$I_{\text{tot}} = I_{\text{direct}} + I_{\text{diffuse}} + I_{\text{reflect}} = 693 + 35 + 138 = 856\ \text{W/m}^2.$$

7.8 SOLAR HEAT GAIN THROUGH FENESTRATION

Fenestration is an opening in a building that is covered by glass or other glazing material. It allows a portion of the incident radiation to pass through, into the building. The most common fenestration are windows and skylights, with glass doors also common for special purposes, e.g., sunrooms. Unlike walls and roofs that have usually significant thermal masses, the thermal mass for fenestration can typically be neglected in energy calculations.

Consider a monochromatic ray on the outer side of a homogeneous fenestration, as shown in Fig. 7.12. The total monochromatic transmissivity,

$$\tau_\lambda = (1-r)^2\, a_\lambda + r^2\,(1-r)^2\, a_\lambda^3 + r^4\,(1-r)^2\, a_\lambda^5 + ..., \tag{7.46}$$

where r is the fraction reflected and a_λ is the fraction available after absorption. This can be simplified into

$$\tau_\lambda = \left[(1-r)^2\, a_\lambda\right] / \left[1 - r^2\, a_\lambda^2\right]. \tag{7.47}$$

Similarly, the total monochromatic reflectivity,

$$\rho_\lambda = r + \left[r\,(1-1)^2\, a_\lambda\right] / \left[1 - r^2\, a_\lambda^2\right]. \tag{7.48}$$

Noting that from

$$\alpha_\lambda + \rho_\lambda + \tau_\lambda = 1, \tag{7.49}$$

the total monochromatic absorptivity,

$$\alpha_\lambda = 1 - \rho_\lambda - \tau_\lambda. \tag{7.50}$$

Therefore,

$$\alpha_\lambda = 1 - r - \left[(1-r)^2\, a_\lambda\right] / \left[1 - r\, a_\lambda\right]. \tag{7.51}$$

The fraction available after absorption, the absorption coefficient,

$$a_{coef} = \exp\left[-KL / \sqrt{\left(1 - \sin^2\theta / n_{ref}^2\right)}\right]. \tag{7.52}$$

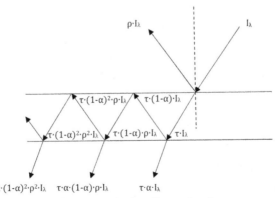

Figure 7.12. Monochromatic transmission through a homogeneous fenestration (created by X. Wang).

Here, K is the extinction coefficient, L is the thickness of the glass, θ is the angle of incidence, n_{ref} is the index of refraction, $n_{ref} = 1.526$, typically.

Energy Exchange through a Glass Window

The rate of **Interior Heat Gain** through the glass can be expressed as

$$q_i' = F_s \tau_{direct} I_{direct} + \tau_{diffuse} I_{diffuse} + \tau_{reflected} I_{reflected} + h_i \left(T_{glass, i} - T_i \right). \quad (7.53)$$

For a single-layer double-strength sheet (DSA) glass, this may be simplified into

$$q_i' = \tau_{avg} I + N_i \alpha_{avg} I + U (T_o - T_i), \quad (7.54)$$

where the fraction of the absorbed radiation becomes a heat gain on the inside,

$$N_i = h_i / (h_i + h_o). \quad (7.55)$$

The fraction of incident solar radiation through a window, both directly transmitted and absorbed and then released inward, can be signified as the **Solar Heat-Gain Coefficient** factor, SHGC. With this, we can write

$$q_i' = SHGC + U (T_o - T_i). \quad (7.56)$$

The tabulated values of SHGC (see Table 10, Chapter 15, ASHRAE 2017 Handbook: Fundamentals [2017]), assume the following:

1. Clear day with a clearness number of 1.0.
2. Ground reflectance, $\rho_{ground} = 0.2$.
3. No external or internal shading.
4. Solar optical properties of DSA glass, where the angle of solar incidence is 60°.
5. Interior and exterior heat transfer film coefficients of 8.3 W/m²·°C and 22.7 W/m²·°C, respectively. These give $N_i = 0.267$.

Deviation from the reference DSA glass can be accounted via the shading coefficient (SC), which is defined as the solar heat gain of the fenestration of concern over the solar heat gain of DSA glass. The SC is used to account for variations in the following:

1. Multiple glazing.
2. Type of glazing materials.
3. Surface coatings.
4. Indoor shading devices.
5. Ground reflectance, $\rho_{ground} \neq 0.2$.

Specifically,

$$q'_i = SC\left[F_s\, SHGC_{direct} + SHGC_{diffuse} + SHGC_{reflected,\,corrected}\right] + U\left(T_o - T_i\right),$$

(7.57)

where

$$SHGC_{reflected,\,corrected} \approx SHGC_{table} + 0.5\, SHGC_{horizontal}\left(\rho_{ground} - 0.2\right).$$ (7.58)

7.9 THE SUN AND THE CYCLE OF LIFE

There is some amount of truth in equating solar energy to life, for there would be no life on planet earth if it were not sustained by solar energy. Figure 7.13 illustrates that the sun provides the needed energy to produce greens[1] (vegetation) via photosynthesis. As such, plants are called autotrophs (self-feeders), because they make their own food from solar energy, water, and carbon dioxide. In turn, herbivores feed on the greens to have energy to thrive. Higher up in the food chain are the omnivores who receive their energy from both plants and animals. Carnivores claim the throne of the food chain by preying on fellow creatures. Nonetheless, even the ferocious carnivores have to fall victim to detritivores, the decomposers that decompose the dead-but-meaty creatures into nutritious soil. The cycle of life

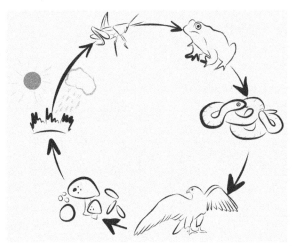

Figure 7.13.   The sun and the cycle of life (created by S. Akhand).

---

[1]Other than plants, there are chlorophyll-bearing organisms such as cyanobacteria, which can capture solar energy and transform it into energy-rich compounds, via photosynthesis, that support the sustenance of other living organisms.

is thus closed, as it takes nutrients in the soil to nourish the plants. Both omnivores and carnivores are heterotrophs (other-feeders), as they are unable to make their own food. As such, whether you are an autotroph, heterotroph, or decomposer, you get your energy from the sun. Someone has graciously put the sun there for us to savor, tapping into solar energy naturally is possibly the most sustainable way to move forward.

PROBLEMS

Problem 7.1

When is the earth farthest away from the sun, where the solar radiation is reduced by 3.3%?

Problem 7.2

On July 1, the horizontal projection of the sun's rays is normal to the west-facing surface of a building located at 45° north of the equator. What is the sun's altitude angle, β?

Problem 7.3

Compare the maximum sun's altitude angle, β_{max}, for a location at (a) the equator, (b) 23.5° north, (c) 45° north, and (d) 45° south.

Problem 7.4

For a west-facing vertical wall of a building in Singapore at 16:00 hour on July 21, what is the extraterrestrial irradiance that would be incident on the wall?

Problem 7.5

At 36° north and 80° west at 11:00 a.m. daylight saving time on June 15, what is the solar angle of incidence for (a) a horizontal surface, (b) a south-facing vertical surface, and (c) a surface tilted 25° from the horizontal and facing 30° south of east?

Problem 7.6

A south-west-facing window on a building in Seoul at 16:00 clock time on July 21 with a height a = 2 m has right at its top a long (large g), horizontal overhang of projection, f = 0.5 m. What is the fraction of the window that is shaded?

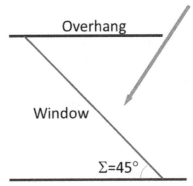

Figure 7.14. An overhang on top of a window with a 45° tilt angle (created by D. Ting).

Problem 7.7

A 2-m wide and 1-m tall south-facing window is to be completely shaded at 10:00 a.m. solar time on June 1, and completely sunlit at 10:00 a.m. solar time on December 1. Find the width of the overhang and its position above the top of the window for (a) Tianjin, (b) Baghdad, (c) Hanoi, Vietnam, (d) Porto, Portugal, and (e) Windsor, Ontario, Canada.

Problem 7.8

A building in Windsor, Nova Scotia (45°N, 64°W), has a south-facing window which is tilted 45° from the horizontal; see Fig. 7.14. The window has a horizontal overhang extending from the base of the window to its top. The ground reflectance, $\rho_g = 0.3$. Find the rate of solar energy incident on the window per unit window area at solar noon on a standard clear day in October.

Problem 7.9

What is the sunlit fraction of a west-facing, 0.1-m wide by 2.5-m high window with a 0.36-m setback in Windsor, Ontario, Canada, at 3:00 p.m. clock time on June 7?

Problem 7.10

A vertical (normal to the ground) stick of 1-m height casts a 1.4-m shadow on the ground in Windsor, Ontario, Canada (42° North, 83° West), on February 22 in the afternoon. What is the solar time? What is the clock time? What direction is the shadow pointing?

Problem 7.11

A south-facing, 2-m high window located at 40°N and 30°E is to be completely shaded at solar noon on summer solstice, and completely sunlit at solar noon on winter solstice. Find the required length of the overhang from the wall, f, and its distance above the top of the window, e (see Fig. 7.7).

REFERENCES

ASHRAE 2017 Handbook: Fundamentals, 2017.

ASHRAE 1993 Handbook: Fundamentals, SI Edition, 1993.

F.C. McQuiston, J.D. Parker, J.D. Spitler, Heating, Ventilation, and Air Conditioning: Analysis and Design, *6th ed., Wiley, Hoboken*, (2005).

O.L. Hadley, T.W. Kirchstetter, "Black-carbon reduction of snow albedo," Nature Climate Change, 2: 437–440, 2012.

T.H. Kuehn, J.W. Ramsey, J.L. Threlkeld, *Thermal Environmental Engineering*, 3rd ed., Prentice-Hall, Upper Saddle River, 1998.

Y. Qian, W.I. Gustafson, Jr., R. Leung, S.J. Ghan, "Effects of soot-induced albedo change on snowpack and hydrological cycle in western United States based on weather research and forecasting chemistry and regional climate simulations," Journal of Geophysical Research, 114: D03108, 2009. doi:10.1029/2008JD011039

CHAPTER 8

Cooling in the Summer

"What good is the warmth of summer, without the cold of winter to give it sweetness."

–John Steinbeck

Nomenclature

A Area; A_{glaz} is the area of the glazing, A_k is the area of the kth element (piece of building envelope), A_{roof} is the area of the roof, A_{wall} is the area of the wall

a Coefficient; $a_0, a_1, a_2, ...a_n$ are coefficients associated with the responses of the transfer function

(Continued)

ACH Air change per hour

b Coefficient; b_0, b_1, b_2, ... b_m are coefficients associated with the driving terms of the transfer function

C Thermal capacitance; C_n is the heat capacity of node n

c Coefficient; c_n are transfer function coefficients of heat conduction through the building envelope

c_p Heat capacity at constant pressure; c_{p_j} is the heat capacity of material or layer j

CLTD Cooling load temperature difference

d Coefficient; d_n are transfer function coefficients of heat conduction through the building envelope

DSA Double-strength A-quality (glass)

F View factor; $F_{w,j}$ is the view factor of the wall with respect to surface j

f Forcing or driving term

g Gravity

h Heat transfer coefficient; h_{air} is the (convection) heat transfer coefficient of air, h_o is the exterior heat transfer coefficient, $h_{o,\,convection}$ is the outdoor convection heat transfer coefficient, $h_{o,\,radiation}$ is the linearized infrared radiative heat-transfer coefficient

h_{fg} Latent heat of vaporization

h_i Enthalpy of indoor air

h_o Enthalpy of outdoor air

HVAC Heating, ventilation, and air conditioning

I Solar irradiance; $I_{diffuse}$ is the diffuse solar radiation, I_{direct} is the direct solar radiation, $I_{reflected}$ is the reflected solar radiation, I_{tot} is the total solar radiation

ICL Instantaneous cooling load

IHG Instantaneous heat gain

K_{tot} Total heat transmission coefficient of the building

k Thermal conductivity; k_j is the thermal conductivity of material or layer j

k_{AC} Air exchange heat transmission coefficient

k_{cond} Total conductive heat transmission coefficient

l Latitude

m Mass; m' is the mass flow rate, m_i' is the mass flow rate of indoor air, m_o' is the mass flow rate of outdoor air

N Number, or, number of hours

(Continued)

n (Node) number

Q Heat; Q' is the heat transfer rate, Q_{cond}' is the conduction heat transfer rate, Q_{ij}' is the heat flow rate from i to j, Q_l' is the latent heat transfer rate, Q_s' is the sensible heat transfer rate

q' Heat flux; $q_{\text{ambient air convection}}'$ is the heat flux because of ambient air convection, $q_{\text{net infrared radiation}}'$ is the heat flux due net infrared radiation, q_o' is the incoming heat flux through the exterior surface, $q_{\text{solar radiation}}'$ is the heat flux because of solar radiation

R Thermal resistance; R_i is the indoor (film) resistance, $R_{i,cov}$ is the inner convection heat transfer resistance, R_{ij} is the resistance between nodes i and j, $R_{i,rad}$ is the inner radiation heat transfer resistance, R_o is the outdoor (film) resistance, $R_{o,conv}$ is the outer convection heat transfer resistance, $R_{o,rad}$ is the outer radiation heat transfer resistance, R_{tot} is the total heat transmission resistance

SC   Solar heat gain of fenestration / solar heat gain of DSA glass

SCL   Solar cooling load factor

SHGF   Solar heat-gain factor

SI   International System of Units

t Time

T Temperature; T_i is the indoor temperature, $T_{i,rad}$ is the inner temperature involved in radiation heat exchange, T_j is the temperature of surface j, T_o is the temperature of the outdoor, $T_{o,conv}$ is the outer temperature that is involved in convection, $T_{o,rad}$ is the outer temperature involved in radiation heat exchange, T_{sa} is the sol–air temperature, T_{wall} is the wall temperature

T' Change in temperature with respect to time; T_n' is the change in temperature of node n with respect to time

TETD   Total equivalent temperature difference

U Heat conductance; U_{glaz} is the U-value of the glazing, U_k is the U-value of the kth element (piece of building envelope), U_{roof} is the U-value of the roof, U_{wall} is the U-value of the wall

V Velocity; V_i is the velocity of indoor air, V_o is the velocity of outdoor air

v Coefficient; $v_0, v_1, v_2, ..., v_n$ are coefficients associated with the transfer function for the IHG terms

W Work; W' is the work output rate

(Continued)

$\boldsymbol{w}$ Coefficient; $w_1, w_2, ..., w_m$ are coefficients associated with transfer function for the ICL terms

$\boldsymbol{x}$ Distance in the x-direction; Δx is the thickness, Δx_i is the thickness of layer i

$\boldsymbol{y}$ Response

Greek and Other Symbols

α Solar absorptivity, or, thermal diffusivity; α_j is the thermal diffusivity of material or layer j

$\boldsymbol{\Delta R}$ Correction; $\varepsilon\Delta R$ accounts for the small-to-moderate differences in the various involved surface temperatures

δ Time lag in hours

ζ Damping magnitude

ρ Density; ρ_j is the density of material j

σ Stefan–Boltzmann constant, $\sigma = 5.669 \times 10^{-8}$ W/(m² · K⁴)

ε Emissivity; ε_j is emissivity of surface j, ε_{wall} is the emissivity of the wall

Σ Tilt angle (also used as a mathematical summation sign, in context)

τ Time, or, time constant

φ_{max} Maximum relative humidity

ω Humidity ratio; ω_i is the indoor air humidity ratio, ω_o is the outdoor air humidity ratio

$\forall$ Volume; $\forall$' is the volume flow rate

8.1 COOLING CHALLENGES AND TERMINOLOGIES

Heating is most needed during cold winter nights, between midnight and dawn, where the sun is by-and-large a non-player and other heat gains are small. Incidentally, we generate much less entropy and heat when we are sound asleep. In contrast, cooling requirement peaks in the summer afternoons, between 1 p.m. and 3 p.m., when the hot sun is high. By this very nature, cooling is further complicated by (1) the varying sun, (2) internal heat gains, and (3) dehumidification.

1) *The varying sun.* Our unceasing rotation and revolution around the sun leads to incessantly varying solar rays, both in terms of direction and intensity, onto and through the building envelope. It is not just the continuous changes associated with the sun, but also the further complexities brought about by the interactions between solar rays and building materials, i.e., significant delay and attenuation caused by thermal storage.

2) *Internal heat gains.* The internal heat gains from indoor sources such as occupants, lights, equipment, and appliances need to be properly accounted for. In general, they do not contribute to any notable challenge during the heating season. This is because the indoor-outdoor temperature difference during the winter is typically much larger than that during the summer time. Also, internal heat gains are acceptable as free energy that helps in lowering the required heating during the cold season. When something is free, it is commonly taken for granted. During the hot summer, however, internal heat gains can add weighty stress on the cooling system, resulting in serious occupant discomfort when they go beyond the rated capacity that the system can handle.

As illustrated in Fig. 8.1, one main challenge with summer cooling is associated with the fact that

Instantaneous Heat Gain $\neq$ Instantaneous Cooling Requirement.

To appreciate this, we need to differentiate heat gain, cooling load, and heat extraction rate. As portrayed in Fig. 8.2,

– Heat gain = thermal energy transferred to and generated within a space.
– Cooling load = thermal energy and humidity removal rates from a space to maintain thermal comfort.
– Heat extraction rate = thermal energy and humidity removal rates by the cooling and dehumidifying equipment.

3) *Dehumidication.* When cooling hot and humid air, we naturally have to invoke some sort of dehumidification. Other than in arid regions, hot summer air retains a lot more moisture than cold winter air. As a consequence, any significant lowering of the air temperature forces the water out of the air, into liquid water[1]. If HVAC engineers do not give the condensed water due attention, they would have to face a wide range of headaches such as mold and material and structural damage. In the winter, sensibly heating cold air can rapidly lower the relative humidity to uncomfortable conditions. With an increase in temperature, the air becomes more thirsty, snatching water from anywhere it can find to quench its thirst. Simply humidifying the dry, warm air to around 50% relative humidity can alleviate problems including eczema, asthma, and nose bleeds. Comparatively, dehumidification is more

---

[1]Recall that air loses its ability to retain moisture as its temperature drops.

Figure 8.1. The heat gain is typically not equal to the cooling requirement at any instant: (a) the cup is gaining heat from the incoming hot water much faster than it loses to the surroundings (created by Y. Yang, edited by D. Ting) and (b) a person almost having a heatstroke is trying to be chilled by air conditioning and cold water, neither seems to be available when needed (created by T.A. Tirtha).

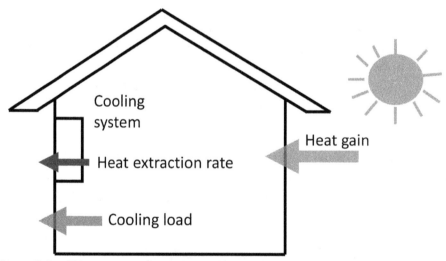

Figure 8.2. When an exterior wall is being heated up by the rising sun, the interior cooling load is less than the exterior heat gain, and the heat extraction rate by the cooling system tends to lag behind the needed cooling load (created by Y. Yang, edited by D. Ting).

challenging as a greater number of parameters need to be taken into consideration. Typically after dehumidification, the air is near saturation. To lower the cool, but muggy, air to the comfortable relative humidity of 50–70% requires some care.

8.2 DESIGN APPROXIMATIONS: WIND SPEED, EXTERIOR CONVECTION, SOL–AIR TEMPERATURE

In accordance with ASHRAE, see Chapter 17 of the *2017 ASHRAE Handbook* [ASHRAE, 2017], the typical indoor design conditions are $T_{dry} = 24°C$ and $\varphi_{max} =$ 50–65%. Even though Chapter 18 of the *2017 ASHRAE Handbook* [ASHRAE, 2017] focuses on non-residential buildings, much of the information is also useful and applicable for residential cooling and heating load estimations.

As a first approximation, the design wind speed for exterior heat convection can be assumed to be caused by 3.4 m/s or 7.5 mph wind, which is one-half the winter value. The corresponding exterior surface heat transfer coefficient, which encompasses both radiation and convection, is $h_{summer} = 22.7$ W/(m²·K), as compared to the winter value, $h_{winter} = 34.0$ W/(m²·K). As mentioned earlier, in Chapter 6, many more details, wind direction, T_{dry}, T_{wet}, etc., have recently been compiled and made available in the CD-ROM of the *2017 ASHRAE Handbook* [ASHRAE, 2017] and other resources. The 1% design T_{dry} and mean coincidence T_{wet} from the climate data in Chapter 14 [ASHRAE, 2017] are generally applicable.

8.2.1 Sol–Air Temperature

Sol–air temperature is a fictitious temperature used to lump together the effects of solar radiation, convection, and infrared radiation into one outdoor temperature that exchanges heat with the exterior surface in question. Specifically, the surface will exchange the same net amount of energy to air at the sol–air temperature as is exchanged in the actual environment. The sol–air approach is being replaced with direct calculations based on factual, detailed climate data monitored by ever-sophisticated instrumentation covering progressively more locations. Nonetheless, going through the exercise with sol–air temperature furnishes understanding of the underlying heat transfer mechanisms involved.

Consider the heat flux onto an exterior wall:

$$q' = Q'/A = q'_{solar\ radiation} + q'_{ambient\ air\ convection} + q'_{net\ infrared\ radiation}, \qquad (8.1)$$

where the solar heat flux,

$$q'_{solar\ radiation} = \alpha\left(I_{direct} + I_{diffuse} + I_{reflected}\right) = \alpha\, I_{tot}. \qquad (8.2)$$

Here, it is assumed that the solar absorptivity, α, is equal for all three components, direct solar radiation, diffused solar radiation from the sky, and reflected solar radiation from the ground and nearby structures. The heat flux of the exterior infrared radiation to the concerned outside wall,

$$q_{\text{net infrared radiation}}{}' = \sum_{j=1}^{j=n} \left(\varepsilon_j \, F_{w,j} \, \sigma \, T_j^4 \right) - \left(\varepsilon_{\text{wall}} \, \sigma \, T_{\text{wall}}^4 \right). \tag{8.3}$$

The first term on the right-hand side signifies the radiation from all the surfaces that the concerned wall can see. Let us assume the emissivity of all involved surface, $\varepsilon_j = \varepsilon_{\text{wall}} = \varepsilon$, a reasonable assumption for non-metal surfaces where $0.9 \le \varepsilon \le 0.95$. Further, consider all surfaces, except the one in question, are at "outdoor" temperature, T_o. To effect this, we use a correction factor $\varepsilon \Delta R$ to account for the small-to-moderate differences in the various involved surface temperatures. Then,:

$$q_{\text{net infrared radiation}}{}' = h_{o,\text{ radiation}} \left(T_o - T_{\text{wall}} \right) - \varepsilon \Delta R, \tag{8.4}$$

where $h_{o,\text{ radiation}}$ is the linearized infrared radiative heat-transfer coefficient, see Fig. 8.3.

The hot outdoor induces a convection heat flux onto the exterior wall:

$$q_{\text{ambient air convection}}{}' = h_{\text{air}} \left(T_o - T_{\text{wall}} \right). \tag{8.5}$$

For practical purposes, we combine the exterior convection heat transfer coefficient with the corresponding radiation coefficient, forming the **exterior film coefficient**:

$$h_o = h_{o,\text{ convection}} + h_{o,\text{ radiation}}. \tag{8.6}$$

In the absence of regional data, or, as a first approximation, the value of this exterior film coefficient may be assumed from the generic design values discussed earlier (second paragraph) in Section 8.2.

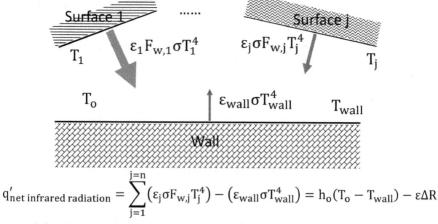

$$q_{\text{net infrared radiation}}' = \sum_{j=1}^{j=n} \left(\varepsilon_j \sigma F_{w,j} T_j^4 \right) - \left(\varepsilon_{\text{wall}} \sigma T_{\text{wall}}^4 \right) = h_o (T_o - T_{\text{wall}}) - \varepsilon \Delta R$$

Figure 8.3. Equivalent radiation from surfaces at dissimilar temperatures via correction factor, $\varepsilon \Delta R$ (created by Y. Yang, edited by D. Ting).

The fallout from the above is a simplified expression for Eq. 8.1, i.e., the total heat flux on the exterior wall in question:

$$q' = \alpha I_{tot} + h_o (T_o - T_{wall}) - \varepsilon \Delta R. \tag{8.7}$$

Introducing the sol–air temperature:

$$T_{sa} = T_o + \alpha I_{tot}/h_o - \varepsilon \Delta R/h_o, \tag{8.8}$$

we have

$$q' = h_o (T_{sa} - T_{wall}). \tag{8.9}$$

This is depicted in Fig. 8.4.

In practice, we assume $\varepsilon \Delta R/h_o$ varies from 0 for vertical surfaces to 4°C (7°F) for upward-facing (horizontal) surfaces. Noting that we are talking about summer, and therefore, the sky overhead is normally colder than the rest of the environment. For tilted surfaces,

$$\varepsilon \Delta R/h_o = 4°C \cos \Sigma, \tag{8.10}$$

where Σ is the tilt angle.

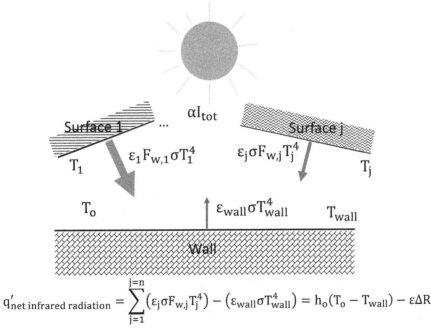

Figure 8.4. Total heat flux onto an exterior surface based on sol–air temperature, T_{sa} (created by Y. Yang, edited by D. Ting).

The ratio, α/h_o, in Eq. 8.8, is a function of the color of the surface. Dark colors have solar absorptivity approaching one, and lighter colors have lower absorptivity values around 0.45, unless the surface is reflective, in which case it can be much lower. This is experientially appreciated on a sunny summer day. Namely, a black shirt under the summer sun feels like a frying pan, and this is not so when wearing a lighter color shirt. For simplicity, we group the exterior surfaces into two categories, dark surfaces with an absorptivity of 0.9 and light surfaces with absorptivity of 0.45, i.e.,

$$\alpha/h_o = 0.9/3 = 0.3 \, \text{hr} \cdot \text{ft}^2 \cdot {}°\text{F/Btu} \, \left(0.052 \, \text{m}^2 \cdot {}°\text{C/W}\right) \text{ for dark-colored surfaces;}$$
$$(8.11)$$

$$\alpha/h_o = 0.45/3 = 0.15 \, \text{hr} \cdot \text{ft}^2 \cdot {}°\text{F/Btu} \, \left(0.026 \, \text{m}^2 \cdot {}°\text{C/W}\right)$$
$$\text{for light-colored surfaces} \qquad (8.12)$$

See Table 15.3 of Kuehn et al. [1998] for sol–air temperatures specifically for a clear day on July 21, $40°$ north latitude, with a maximum T_{dry} of 35°C (95°F) and a daily range of 11.7°C (21°F).

EXAMPLE 8.1. SOL–AIR TEMPERATURE FOR LIGHT AND DARK SURFACES

Given: A vertical surface with incident solar radiation of 200 Btu/hr·ft², where T_o is 80°F.

Find: The sol–air temperature difference between a light and dark surface.

Solution:

The sol–air temperature according to Eq. 8.8 is

$$T_{sa} = T_o + \alpha \, I_{tot}/h_o - \varepsilon \, \Delta R/h_o.$$

The last term is the correction factor for reflected radiation from surfaces at dissimilar temperatures, divided by the exterior heat transfer coefficient. For non-vertical surfaces, we have

$$\varepsilon \, \Delta R/h_o = 7°\text{F} \cos \Sigma,$$

and

$$\alpha/h_o = 0.3 \, \text{hr} \cdot \text{ft}^2 \cdot {}°\text{F/Btu for dark-colored surfaces,}$$

$$\alpha/h_o = 0.15 \, \text{hr} \cdot \text{ft}^2 \cdot {}^\circ\text{F/Btu for light-colored surfaces.}$$

Therefore, for dark-colored vertical surfaces,

$$T_{sa} = 80 + 0.3\,(200) - 70\,(0) = 140{}^\circ\text{F},$$

and

$$T_{sa} = 80 + 0.15\,(200) - 70\,(0) = 110{}^\circ\text{F}.$$

We see that lightening the exterior walls not only brightens the building outlook but also can lead to significant savings in air conditioning.

8.3 PERIODIC HEAT GAIN THROUGH WALLS AND ROOFS

The heat gain through walls and roofs is periodic, as the outdoor conditions associated with the sun rising and setting are periodic (cyclic) in nature. The relatively bulky building elements such as the walls and the roof act like a buffer with a finite thermal mass (capacitance). Consequently, the walls and the roof bring forth delay and attenuation, as the outdoor environmental condition cycles through them.

Consider one-dimensional transient heat transfer through a multilayered walls as shown in Fig. 8.5. The change in temperature with respect to time of layer j is

$$\partial T/\partial t = \alpha_j \, \partial^2 T/\partial x^2, \tag{8.13}$$

where $\alpha_j = k_j/\rho_j c_{p,j}$, the thermal diffusivity of the jth layer. Recall that the thermal diffusivity represents how fast heat propagates through the material, i.e., the

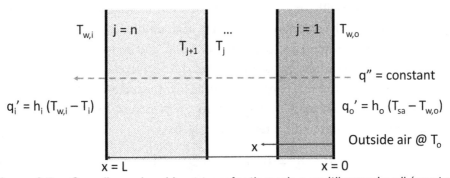

Figure 8.5. One-dimensional heat transfer through a multilayered wall (created by D. Ting). Perfect contact between consecutive layers has been assumed.

amount that passes through with respect to the amount that is being absorbed by the material as it passes through. For practical purposes, it is assumed to be constant. At the interface between two adjacent layers, the heat flux and temperature must be equal in both materials under steady-state condition, i.e.,

$$k_j \left(\partial T / \partial x \right)_j = k_{j+1} \left(\partial T / \partial x \right)_{j+1}, \tag{8.14}$$

and

$$T_j = T_{j+1}. \tag{8.15}$$

The boundary condition on the indoor or interior surface is

$$q_i' = h_i \left(T_{w,i} - T_i \right). \tag{8.16}$$

In other words, the heat flux from the interior wall to the indoor air is proportional to the temperature difference between the wall and the air. The proportionality constant is the indoor heat transfer coefficient, h_i. In the summer, the heat is transmitted from the warmer outdoors to the cooler indoors. It follows that the heat flux into the indoor air via convection is equal to that conducted through the inner surface, i.e.,

$$q_i' = h_i \left(T_{w,i} - T_i \right) = -k_n \left(\partial T / \partial x \right)_{x=L}. \tag{8.17}$$

Similarly, on the outdoor side, the heat flux into the exterior surface is

$$q_o' = -k_1 \left(\partial T / \partial x \right)_{x=0} = h_o \left(T_{sa} - T_{w,o} \right), \tag{8.18}$$

where the heat flux is positive when the energy is moving into the building.

The above equations can be solved to express the temperature, T, at any location, x, as a function of the various involved parameters. As the outdoor parameters, T_{sa} in particular, change with time, the incoming heat flux through the exterior surface, q_o', also fluctuates with respect to time. For a typical summer day, T_{sa}, and hence, q_o' can be described by a sine wave, peaking near 3 p.m. and bottoming around 3 a.m.

As the exterior heat flux, q_o', is transmitted through the layers, it is attenuated because some of the thermal energy is absorbed by the materials. Furthermore, it takes time for the heat flux to propagate through the layers. Resultantly, the IHG (the rate of heat transfer to the interior), see Fig. 8.6,

$$q_i' = \text{IHG} = h_i \left(T_{w,i} - T_i \right), \tag{8.19}$$

lags behind q_o' and it fluctuates at a lower amplitude.

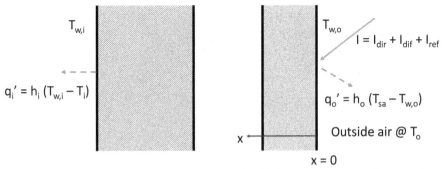

Figure 8.6. The instantaneous heat gain (IHG) is the heat transfer rate to the interior space (created by D. Ting). IHG is the heat entering the indoor air; it is almost never equal to the IHG entering the exterior wall.

It is clear that changes in the building envelope composition and dimension have a direct influence on the resulting dynamic behavior of the layered envelope of concern. The solution of the IHG thus varies from wall to wall, wall to roof, and from building to building. The most direct and potent solution is empirical, where the dynamic behavior of sections of the building envelope is quantified in a controlled environment. These tests, complimented by analytical solutions, generate a set of values associated with different parameters involved. A couple of common solutions from this kind of test are given below.

8.3.1 Transfer-Function Method

The instantaneous (cooling) load can be considered as the response of the building to the driving terms, such as q_o, T_o, and T_i, that act on it. Suppose there is a single driving (forcing) term, $f = f(t)$, and the response is $y = y(t)$. According to the transfer function model, the relation between the response and the driving term can be expressed as

$$y_t = -\left(a_1 y_{t-\Delta t} + a_2 y_{t-2\Delta t} + \cdots + a_n y_{t-n\Delta t}\right)$$
$$+ \left(b_0 f_t + b_1 f_{t-\Delta t} + b_2 f_{t-2\Delta t} + \cdots + b_m f_{t-m\Delta t}\right),$$
(8.20)

where the time step, Δt, is generally taken as 1 hr, and a_1 to a_n and b_0 to b_m are coefficients that characterize the system, and they are assumed to be independent of f_t or y_t. The influence of thermal inertia is accounted for through these coefficients. The values of these coefficients diminish rapidly as we move further back into the past. This implies that the current response, y_t, is most influenced by the most recent forcing, f_t and $f_{t-\Delta t}$, and much less so by what happened in the more distant past. Similarly, current response, y_t, also depends more on the most recent response, $y_{t-\Delta t}$, and significantly less on distant-past responses. Mathematically, we can introduce a_0 and recast Eq. 8.20 as

$$
\begin{aligned}
a_0\, y_t + a_1\, y_{t-\Delta t} + a_2\, y_{t-2\Delta t} + &\ldots + a_n\, y_{t-n\Delta t} \\
= b_0\, f_t + b_1\, f_{t-\Delta t} + b_2\, f_{t-2\Delta t} + &\ldots + b_m\, f_{t-m\Delta t}.
\end{aligned}
\tag{8.21}
$$

To bring the above into context, let us consider the response as the heat input rate, Q', which is the rate of heat entering the indoors. This response can be deduced from two driving terms, indoor temperature, T_i, and outdoor temperature, T_o. As such, we can write

$$
\begin{aligned}
a_{Q,0}\, Q_t' + a_{Q,1}\, Q_{t-\Delta t}' + a_{Q,2}\, Q_{t-2\Delta t}' + &\ldots + a_{Q,n}\, Q_{t-n\Delta t}' \\
= b_{i,0}\, T_{i,t} + b_{i,1}\, T_{i,t-\Delta t} + &\ldots + b_{i,m}\, T_{i,t-m\Delta t} \\
+ b_{o,0}\, T_{o,t} + b_{o,1}\, T_{o,t-\Delta t} + &\ldots + b_{o,p}\, T_{o,t-p\Delta t}.
\end{aligned}
\tag{8.22}
$$

Physically, the equation asserts that the heat input rate is influenced by the heat input rate at an earlier time. This can be visualized as the heating up of the building structure under consideration by heat input prior to the instant of interest. Thereupon, the present heat input rate is also affected, in a diminishing manner, by both indoor and outdoor temperatures earlier. The equation can be used for solving any one of the three variables, Q_t', $T_{i,t}$, and $T_{o,t}$, hour by hour, provided two of the three variables are known and that the coefficients, a_Q, b_i, and b_o, have been deduced from measurements.

We see that the same transfer function method can be used to relate the interior heat flux (from indoor surface into indoor air), q_i', with the exterior heat flux (on the outer surface of the building), q_o'. This is shown in Fig. 8.7. Figure 8.7(a) shows that because of the finite thermal resistance, or less than infinite thermal conductance, U, of the building envelope, not all the heat makes it through the building envelope. As a result, q_i' is less than q_o' when the outdoor (equivalent) temperature is higher than the design temperature, $T_{i,design}$, at which temperature neither cooling nor heating is needed. When the outdoor (equivalent) temperature drops below the design temperature, q_i' does not drop as much as q_o'. As such, less heating is required. With the finite thermal mass of the building envelope, there is further attenuation, Fig. 8.7(b). As importantly, the thermal mass imposes some delay. The summed effect is depicted in Fig. 8.7(c), where q_i' is largely leveled out compared to, and lags behind, q_o'.

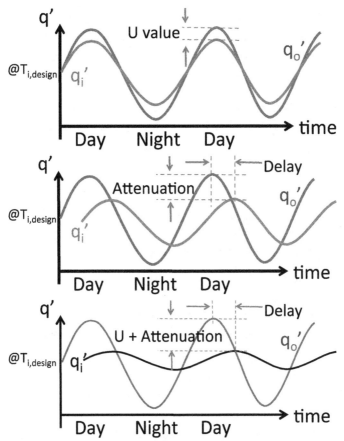

Figure 8.7 Relating the interior heat flux, q_i', with the exterior heat flux, q_o': (a) attenuation because of finite U; (b) delay and attenuation imposed by thermal mass; and (c) overall effect of finite U and thermal mass on q_i' (created by D. Ting).

8.3.2 Instantaneous Cooling Load

It is clear that because of the thermal mass of the building envelope, the indoor IHG is of lower magnitude and lags behind that of the outdoor heat gain. The corresponding cooling needed at any moment is also not equal to the instantaneous indoor heat gain. In other words, part of the IHG released from the indoor surface goes to indoor building structures and furniture. These items have thermal mass, enabling them to store sensible heat. Most items can also retain moisture, and hence, can store latent heat. The portion of energy convected into the indoor air becomes a nearly instantaneous load on a forced-air cooling coil. On the other hand, the infrared-radiation portion of the heat gain is primarily absorbed by and stored in the indoor materials such as building structures and furniture. This stored energy gradually raises the temperature of these materials until they, too, become convective heat sources within the space, but at a later time. The time constants associated with moisture storage are relatively large. Consequently, this latent storage mechanism is often neglected in hourly transient analyses.

Transfer-Function Method for ICL   The transfer-function concept can be applied to relate the ICL with the IHG. The load at any time τ can be described as a response to the current and previous heat gains minus the effect of the previous cooling loads, i.e.,

$$
\begin{aligned}
\mathrm{ICL}_\tau/A = (v_0\,\mathrm{IHG}_\tau + v_1\,\mathrm{IHG}_{\tau-\tau} + v_2\,\mathrm{IHG}_{\tau-2\tau} + ...) \\
- (w_1\,\mathrm{ICL}_{\tau-\tau} + w_2\,\mathrm{ICL}_{\tau-2\tau} + ...).
\end{aligned}
\tag{8.23}
$$

For most practical purposes, this can be truncated as

$$
\mathrm{ICL}_\tau/A = (v_0\,\mathrm{IHG}_\tau + v_1\,\mathrm{IHG}_{\tau-\tau}) - w_1\,\mathrm{ICL}_{\tau-\tau}.
\tag{8.24}
$$

Sample values of coefficients v_0, v_1, and w_1 can be found in Tables 16.1 and 16.2 of Kuehn et al. [1998]. Note that these and other coefficients discussed in this chapter are adopted from ASHRAE handbooks. As ASHRAE handbooks evolved over the years, some of these concepts and coefficients tend to disappear and reemerge, depending predominantly on the handbook committees, whose members and their views tend to change from year to year.

The point is that the values of the aforementioned coefficients, along with the present and past IHGs, must be known before we can apply the method. Furthermore, the following points are also worth highlighting.

1) Because of the thermal storage effect, the maximum instantaneous cooling load, ICL_{max}, is smaller than the IHG_{max}. We see that a large thermal mass can be beneficial. This is particularly the case during the spring and fall seasons, where the daily high requires cooling and the low at night necessitates heating. Uniforming out the high and low can significantly reduce the need of both cooling and heating. Some of the large "ancient" buildings built of rock, soil, or clay furnish a massive thermal mass, allowing them to level out more than daily but seasonal weather fluctuations.

2) It follows from Highlight 1 that the ICL is more uniform than the IHG.

3) When the heat gain goes negative over short time periods, the cooling load tends to remain positive.

4) Simulations typically take a few days to stabilize to a repeatable ICL cycle, as initial guesses may be far away from the solutions.

8.3.3 Total Equivalent Temperature Difference

The transfer-function method discussed above suffers from one major drawback, i.e., the need of a multitude of values for the various coefficients which vary from construction to construction. To ease engineering practice, numerous simpler methods have been sought over the decades. Let us look at the total equivalent temperature difference (TETD) approximation.

Conductive Heat Gain   The conductive heat gain (or loss), $Q_{\text{cond},t}$,' at time t through the roof and walls is

$$\dot{Q}_{\text{cond},t} = -\sum_{n=1} d_n \dot{Q}_{\text{cond},t-n\Delta t} + A\left(\sum_{n=0} b_n T_{\text{sa},t-n\Delta t} - T_i \sum_{n=0} c_n\right), \qquad (8.25)$$

where A is the area of roof or wall in m^2, time step, $\Delta t = 1$ h, $T_{\text{sa},t}$ is the sol–air temperature at time t, and b_n, c_n, and d_n are coefficients of the conduction transfer function. Tables 15.8 and 15.9 of Kuehn et al. [1998] and Tables 7.7–7.9 of Kreider et al. [2002] list sample values of the coefficients of the conduction transfer function. Note that the indoor temperature, T_i, is multiplied by the sum of the c_n values because T_i has been assumed to be constant. In general, the initial value, $Q_{\text{cond},t=0}$, is not known. Therefore, the calculation needs to be repeated over a sufficient number of time steps (a few days) until the resulting pattern becomes periodic within the desired accuracy.

The transfer-function coefficients can be related with the U-value. In the steady-state limit, where Q_{cond},' T_{sa}, and T_i are all constant, we have

$$\dot{Q}_{\text{cond}} \sum_{n=0} d_n = A\left(T_{\text{sa}} \sum_{n=0} b_n - T_i \sum_{n=0} c_n\right), \qquad (8.26)$$

where $d_0 = 1$. Also,

$$Q_{\text{cond}}{}' = A\,U\,(T_{\text{sa}} - T_i), \qquad (8.27)$$

implying

$$\sum b_n = \sum c_n, \qquad (8.28)$$

and therefore,

$$U = \sum c_n / \sum d_n. \qquad (8.29)$$

To simplify, we introduce the TETD that accounts for T_{sa} variation outside and the damping and time lag associated with the construction of the wall or roof. By its very nature, the IHG,

$$q_i' = h_i \left(T_{wall,i} - T_i\right) = U\left(\text{TETD}\right).$$ (8.30)

If q_i' lags the T_{sa} variation by δ hours and the magnitude is damped by ζ, then the total equivalent temperature difference,

$$\text{TETD}_\theta \approx \left(T_{sa,avg} - T_i\right) + \zeta\left(T_{sa,\delta} - T_{sa,avg}\right).$$ (8.31)

8.3.4 Cooling Load Temperature Difference (CLTD) Method

The TETD method had been replaced by the cooling load temperature difference (CLTD) method. Namely, an approximate hand-calculation method for determining summer design cooling load is the ASHRAE CLTD method.

Walls and Roofs We may divide the sensible heat transfer via walls and roofs into convective and radiative parts. The convective portion more-or-less translates itself immediately into the ICL. The radiative part, however, is stored and spread out into an average cooling load over a period.

The ICL at time τ,

$$\text{ICL}_\tau = A\left[q_{\tau,conv}' + \sum_{n=0}^{n=N} q_{\tau-n\Delta\tau}'/N\right],$$ (8.32)

where the radiant portion of the heat gain is averaged over N hours. For hand calculations, we can express ICL as a function of TETD:

$$\text{ICL}_\tau = UA\left[(\% \text{ convection}) \text{TETD} + (\% \text{ radiation}/N) \sum_{n=0}^{n=N} \text{TETD}_{\tau-n\Delta\tau}\right].$$ (8.33)

One may compare this ICL with the IHG expressed by Eq. 8.30. Approximately 60% of the interior surface film coefficient is infrared radiation and 40% convection, i.e., ASHRAE recommends 0.4 and 0.6 for % convection and % radiation in Eq. 8.33, respectively. ASHRAE also recommends $N = 2$–3 h for lightweight construction and 6–8 h for heavy construction. The ICL expression can be recast as

$$\text{ICL}_\tau = U\,A\,\text{CLTD}_\tau,$$ (8.34)

where values for CLTD are tabulated in Tables 16.5 and 16.6 (based on clear-sky weather conditions for July 21, 40°N) of Kuehn et al. [1998]. Also see Tables 15.8–15.10 and 16.7 [Kuehn et al. 1998]. Note that CLTD is also sort of phased out; see Chapter 18, Section 10, or page 18.57, of the *ASHRAE 2017 Handbook* [ASHRAE, 2017].

Fenestrations Recall that the IHG through a fenestration per unit surface area can be estimated from

$$q_i = SC \cdot SHGF + U(T_o - T_i), \qquad (8.35)$$

where SC is the solar heat gain of fenestration/solar heat gain of double-strength A-quality (DSA) glass, SHGF is the solar heat-gain factor. The first term is the interior heat gain from the amount transmitted through the glazing plus the amount absorbed by the glazing material and added to the interior. As the thermal mass of the fenestration is substantially less than the building (structural) thermal mass, it can be omitted.

The cooling load is delayed and reduced in magnitude, owing to the storage of absorbed infrared radiation inside the building. The cooling load can be estimated, for the heat gain $U(T_o - T_i)$, from using CLTD, Table 16.9 of Kuehn et al. [1998]. Therefore,

$$ICL_{\tau,cond} = U \cdot A \cdot CLTD_\tau, \qquad (8.36)$$

where the total area, glazing plus framing, is used with the average U-value for the entire window. The time delay of the first term can be accounted for by introducing the **solar cooling load factor**, SCL.

To repeat, the bulk of these methods is no longer included in the latest ASHRAE handbooks, though they may resurface in future versions. Interested readers can refer to Kuehn et al. [1998], for example. Table 16.11 of Kuehn et al. [1998] tabulates typical values of SCL for $l = 40°$ North in July. ASHRAE recommends north-facing values for shaded areas, as this includes sky-diffuse and ground-reflected radiation but not direct solar radiation. In short, the total ICL through the fenestration is

$$ICL_\tau = A_{glass}\, SC\, SCL_\tau + U\, A\, CLTD_\tau. \qquad (8.37)$$

8.3.5 Internal Heat Gain

Other than heat transmission through the building envelope, heat is also generated indoors. The most common indoor heat generating sources are (1) occupants, (2) lighting, (3) appliances, and (4) infiltration.

1) *Occupants.* The wisecrack about having people over for a dancing party to lower the heating bill in the winter is no joke. An individual can produce in excess of 400 Watts when dancing. Accordingly, a party of 50 active dancers equates to over 20 kW of thermal energy. Unsurprisingly, some windows have to be opened to vent the excess heat so that the party animals do not pass out from heat stroke. Although this type of winter partying seldom happens, every Watt of heat generated by an occupant in the summer has to be removed. As such, the internal heat gain from occupants needs to be duly accounted for in engineering air conditioning. Table 8.1 lists the heat output from a typical adult undergoing different physical activities. The data are generated from the typical metabolic heat generation tabulated in Table 4, Chapter 9 of the *ASHRAE 2017 Handbook* [ASHRAE, 2017]. The flux in W/m² is multiplied by 1.8 m² for an average adult's skin area. Table 1, Chapter 18 of the *ASHRAE 2017 Handbook* [ASHRAE, 2017] furnishes representative rates at which heat and moisture are given off by human beings in different states of activity. The breakdown of the generated heat into sensible and latent components allows better estimation of the portion affecting the ICL immediately versus that which averages out over many hours.

2) *Lighting.* The widespread replacement of heat-intensive incandescent lights with florescent lights is an overdue awakening of the fact that the old style of lighting has a toll on the air-conditioning requirements. LED (light-emitting diode) lights have recently made inroads to further reduce unnecessary heat generation when only illumination is desired. Tables 2 and 3, Chapter 18 of the *ASHRAE 2017 Handbook* [ASHRAE, 2017], detail many lighting power

Table 8.1. Heat-generation rate of an average adult undergoing different activities.

Activity	Sleeping	Seating	Standing	Walking	Dancing
Heat-generation rate [W]	70	110	130	180	250–500

densities. As a first approximation, the wattage specified for each light can be utilized. This, however, tends to overestimate the heat gain rate, as part of the lighting energy goes beyond the confined indoor space. On the other hand, Kuehn et al. [1998] state that a fluorescent light ballast can add 20% more heat gain than the lamp itself. Therefore, using the rated power may not be a bad start. Note that unlike the heat gain from occupants, lighting only gives out sensible heat.

3) *Appliances.* The heat gain from appliances varies as widely as the appliances that we have in this electrical age. Thereupon, Tables 5A–5F, and the associated text in Chapter 18 of the *ASHRAE 2017 Handbook* [ASHRAE, 2017] are devoted to the heat gain from appliances. The rated power of each appliance may be used as a first estimate of the heat gain.

4) *Infiltration.* All buildings have cracks and openings, including doors and windows. Outdoor air infiltrates through them at the location where the indoor pressure is lower than the outdoor pressure, as discussed in Chapter 6. This infiltration is balanced by exfiltration through the apertures where the indoor pressure is higher than that of the outdoors. Although the typically fresher air can be a welcome guest, the guest can be rather sweaty and uncomfortably warm in the middle of the summer. As such, infiltration is considered as an internal heat gain source, where both sensible and latent heat is produced. For this reason, these heat gains put a noticeable demand on the cooling capacity. As humidity is prevailing in the summer outdoor air, it makes sense to deal with internal heat gain because of infiltration by separating the sensible heat from the latent heat. The sensible heat gain,

$$Q_s' = \forall' \rho \, c_p \, (T_o - T_i), \qquad (8.38)$$

where $\forall'$ is the volume flow rate of infiltration, ρ is the density of the infiltrated outdoor air, c_p is the corresponding heat capacity, T_o is the temperature of the outdoor air, and T_i is the indoor temperature. The latent heat gain,

$$Q_l' = \forall' \rho \, h_{fg} \, (\omega_o - \omega_i), \qquad (8.39)$$

where h_{fg} is the latent heat of vaporization, ω_o is the humidity ratio of the outdoor air, and ω_i is the humidity ratio of the indoor air.

8.4 TOTAL COOLING AND HEATING LOADS

The three most important terms in the heat balance of a building are (1) heat flow across the building envelope, (2) air exchange, and (3) indoor heat sources/sinks.

8.4.1 Heat Flow across the Building Envelope

The total conductive heat flow from exterior to interior across the building envelope can be approximated as

$$Q_{cond}' = \sum U_k A_k (T_o - T_i). \tag{8.40}$$

This equation considers each piece of building envelope with a unique U-value separately. Summing the heat across all the pieces of the entire building envelope gives the total conductive heat flow from outdoors into the indoors. The expression can be simplified by introducing the **Total Conductive Heat Transmission Coefficient**, $k_{cond} = \sum U_k A_k$. Namely,

$$Q_{cond}' = k_{cond} (T_o - T_i). \tag{8.41}$$

Conventionally, values of total conductive heat transmission coefficient are tabulated, and used to estimate the heat gain or loss of buildings of similar construction.

The building envelope consists of a large number of different parts which can be divided into three main groups: (1) glazing, (2) opaque walls, and (3) roof. Therefore, the **heat transmission coefficient** can be expressed as

$$k_{cond} = U_{glaz} A_{glaz} + U_{wall} A_{wall} + U_{roof} A_{roof}. \tag{8.42}$$

8.4.2 Heat Flow Because of Air Exchange

Infiltration has been regarded as an internal heat gain in the summer time, as discussed in the previous section. The heat flow because of air exchange, in reality, occurs across the building envelope. Therefore, the associated amount of heat flow can be analyzed as another heat transmission across the building envelope. This **air exchange heat transmission coefficient**,

$$k_{AC} = \rho\, c_p\, \forall'. \tag{8.43}$$

As a first approximation, only the sensible heat needs to be considered.

EXAMPLE 8.2. COOLING NEED CAUSED BY AIR EXCHANGE

Given: A residential building at sea level has a air exchange rate of one. Infiltration is balanced by an equal mass of outgoing flow called exfiltration, i.e., mass conservation under steady conditions. On a summer day, the outdoor air temperature is 35°C. The building has 150 m² of floor area and is two stories high with a net interior height of 7 m.

Find: The amount of cooling needed to compensate for the heat gain because of air exchange to maintain the indoor air temperature at 22°C.

Solution:

For the open system of concern, the first law of thermodynamics (energy conservation under steady state, steady flow conditions) can be expressed as

$$m_o' \left(g z_o + V_o^2/2 + h_o\right) = m_i' \left(g z_i + V_i^2/2 + h_i\right) + Q' + W', \qquad (8.2.1)$$

where subscript 'o' signifies the incoming outside air, subscript 'i' denotes the outgoing inside air, m' is the mass flow rate, g is gravity, z is the elevation, V is the velocity, h_o is enthalpy of outdoor air, h_i is the enthalpy of indoor air, Q' is the rate of heat needs to be removed, and W' is the work output rate, which is zero. Neglecting the inconsequential changes in the potential and kinetic energy, the expression is simplified into

$$m_o' h_o = m_i' h_i + Q'. \qquad (8.2.2)$$

Conservation of mass gives $m_o' = m_i'$, and hence, we have the cooling rate,

$$Q' = m \left(h_o - h_i\right). \qquad (8.2.3)$$

For practical purposes, this can be approximated as

$$Q' = \rho c_p \forall' \left(T_o - T_i\right), \qquad (8.2.4)$$

where $\rho c_p = 1.2$ kJ/(m³·K) at standard atmospheric conditions.

8.4.3 The Total Heat Transmission Coefficient

We can combine the heat transmission coefficient across the building envelope with that associated with air exchange into a **Total Heat Transmission Coefficient**,

$$K_{tot} = k_{cond} + \rho\, c_p\, \forall'. \tag{8.44}$$

Recall that the first term on the right hand side is the total conductive heat transmission coefficient which is used to denote the overall heat flow across a building envelope except that because of air exchange. The second term accounts for the heat flow because of air exchange.

EXAMPLE 8.3. THE TOTAL HEAT TRANSMISSION COEFFICIENT.

Given: A simple wood-frame building is built as a rectangular box, 12 m by 12 m by 2.5 m, with a flat roof. The insulation is fiberglass with $k = 0.06$ W/(m·K) and a thickness of 0.25 m in the roof and of 0.15 m in the walls. The windows are double-glazed with $U_{glaz} = 3.0$ W/(m²·K) and they cover 20% of the sides. The air exchange rate is 0.5 per hour.

Find: The heat transmission coefficient, K_{tot}.

Solution:

For simplicity, treat the opaque surfaces as if they consist only of the fiberglass, neglect thermal bridges because of studs, and assume all values to be constant.

The total heat transmission coefficient according to Eq. 8.44 is

$$K_{tot} = k_{cond} + \rho\, c_p\, \forall'.$$

Compo-nent	Area [m²]	Thick-ness [m]	Conduc-tivity [W/m·K]	U (k/Δx) [W/m²·K]	k (UA, $\rho c_p \forall$)[W/K]
Roof	144	0.25	0.06	0.24	34.6
Walls	96	0.15	0.06	0.40	38.4
Glazing	24			3	72
Air Exchange					60
Total					205

The total heat transmission, K_{tot}, is 205 W/K.

8.4.4 Transient Thermal Network

The entire building may be modeled based on an electrical circuit analogy. An resistance–capacitance network can be used to lump the entire heat capacity of a building into a single massive node. Though crude, this gives the designer a simple tool for estimating the warm-up and cool-down times associated with thermostat setback and setback recovery. Note that setting the thermostat back, to a lower temperature during the heating season and/or setting it up to a higher temperature during the cooling season, when a building is unoccupied can potentially save a huge amount of energy.

We can approximate a building as being composed of a finite number of parts, N, called nodes, each of which is assumed to be isothermal. The heat flow between neighbors is

$$Q_{ij}' = \left(T_i - T_j\right)/R_{ij}. \qquad (8.45)$$

The heat balance of node n is

$$C_n \dot{T}_n = \sum_{i=1}^{N} \frac{T_i - T_n}{R_{in}} + \dot{Q}_n, \qquad (8.46)$$

where C_n is the heat capacity of node n, temperatures T are analogous to voltage, and heat flows Q' are analogous to current.

Consider a single pane window between indoor air at T_i and outdoor air at T_o, with surface heat transfer coefficients h_i and h_o (combining radiation & convection), as depicted in Fig. 8.8. The thermal capacitance of the glass,

$$C = \rho\, c_p\, A\, \Delta x. \qquad (8.47)$$

The inner and outer thermal resistance are, respectively,

$$R_i = 1/\left(A\, h_i\right), \qquad (8.48)$$

and

$$R_o = 1/\left(A\, h_o\right). \qquad (8.49)$$

If there is no absorption of radiation in the glass, $Q' = 0$, then,

$$CT' = (T_i - T)/R_i + (T_o - T)/R_o. \qquad (8.50)$$

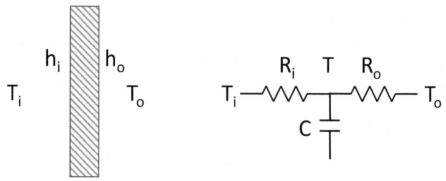

Figure 8.8.   Thermal networks of a single-pane window (created by D. Ting).

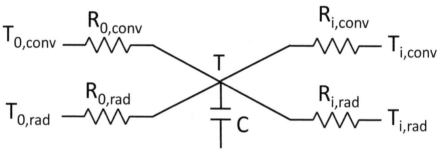

Figure 8.9.   Thermal networks of a single-pane window coupled with radiation and convection at different temperatures, with no radiation absorption (created by Y. Yang).

Radiation and convection are coupled to nodes at different temperatures, therefore, as shown in Fig. 8.9,

$$CT' = (T_{i,rad} - T)/R_{i,rad} + (T_{i,conv} - T)/R_{i,conv}$$
$$+ (T_{o,rad} - T)/R_{o,rad} + (T_{o,conv} - T)/R_{o,conv}. \tag{8.51}$$

Consider a wall composed of N layers each of thickness Δx_i, as sketched in Fig. 8.10. The heat balance of internal nodes is

$$CT_n' = (T_{n-1} - T_n)/R + (T_{n+1} - T_n)/R, \tag{8.52}$$

for $2 \leq n \leq N - 1$. The equations for the external nodes are

$$CT_1' = (T_i - T_1)/R_i + (T_2 - T_1)/R, \tag{8.53}$$

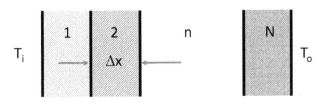

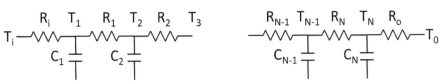

Figure 8.10. Thermal networks of a wall composed of multiple layers (created by D. Ting).

and

$$CT_n' = (T_{n-1} - T_n)/R + (T_o - T_n)/R_o. \tag{8.54}$$

The indoor thermal resistance

$$R_i = R_1/2 + 1/(h_i A). \tag{8.55}$$

The resistance of the first layer, typically drywall,

$$R_1 = \Delta x_1/(k_1 A). \tag{8.56}$$

The outdoor (exterior) thermal resistance,

$$R_o = R_N/2 + 1/(h_o A). \tag{8.57}$$

The thermal capacitance of the first layer from the indoor side,

$$C_1 = \rho_1 c_{p,1} A \Delta x_1. \tag{8.58}$$

Time Constant To model thermal inertia, a network must contain at least one capacitor. For example, let us consider the indoor air temperature of a simple building as a node, as illustrated in Fig. 8.11. Then, the **total heat transmission resistance** of the simplified single-zone building,

$$R_{tot} = 1/K_{tot}, \tag{8.59}$$

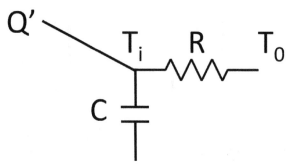

Figure 8.11. Thermal networks of a simplified single-zone building (created by D. Ting). The entire building with a total capacitance of C and a total heat transmission resistance R is at T_i .

where K_{tot} is the total heat transmission coefficient of the building. The heat balance is

$$C\,T_i' = (T_o - T_i)/R + Q'. \tag{8.60}$$

Defining the time constant,

$$\tau = RC, \tag{8.61}$$

then

$$\tau T_i' + T_i = T_o + R\,Q'. \tag{8.62}$$

For constant T_o and Q', and noting that

$$d\,[T(t) = T_i(t) - T_o - R\,Q']/dt = T_i', \tag{8.63}$$

the equation becomes

$$\tau T' + T = 0. \tag{8.64}$$

The solution for this first-order differential equation is

$$T(t) = T(0)\,\exp(-t/\tau). \tag{8.65}$$

We note that after one time constant, $T(t)$ decays to $1/e \approx 0.368$ of its initial value $T(0)$.

A more comprehensive revelation of the underlying elements which make up the total resistance of the single-zone building is exhibited in Fig. 8.12. The problem becomes significantly more complex if we further relax the model to account for multiple zones with different indoor temperatures, T_i. A simple way to deal with this is by considering each zone as a separate entity, exchanging heat with neighboring zones at different temperatures via their respective resistance and capacitance.

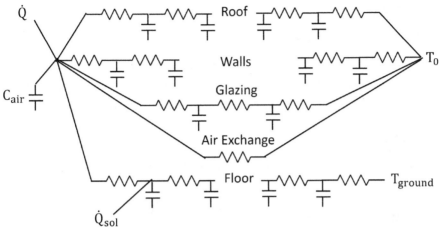

Figure 8.12. Thermal networks of a single-zone building with heat exchanging via the different elements of the building envelope (created by Y. Yang).

EXAMPLE 8.4. TIME CONSTANT OF A BUILDING

Given: Suppose a building can be described by the RC network as shown in Fig. 8.11. T_i, T_o, and the heat input, Q, have been constant until $t=0$, when the thermostat is set back.

Find: The time taken for T_i to drop from 20°C to 15°C, if $R = 5.0$ K/kW, $C = 10$ MJ/K, and $T_o = 0$°C.

Solution:

The exponential drop (decrease) can be described as

$$T(t) = T(0) \exp(-t/\tau)$$

or,

$$15°C = 20°C \exp(-t/\tau)$$

The time constant, $\tau = RC = 5$ K/kW $(10$MJ/K$) = 50{,}000$ s

Substituting this into the equation and solving it leads to

$$t = 14{,}384 \text{ s} = 4 \text{ hr}$$

It will take 4 hours for the indoor temperature to drop from 20°C to 15°C.

PROBLEMS

Problem 8.1

Figure 8.13 shows the instantaneous heat flux, q_o', on the outside wall of a building. Accurately sketch and clearly label (a) the instantaneous heat flux of the inner wall, q_i' (IHG), and (b) the corresponding ICL. Explain the variation of each.

Problem 8.2

The sol–air temperature, T_{sa}, is 47°C. What is the temperature of a perfectly insulated, light-colored wall?

Problem 8.3

An elementary school with an indoor volume of 2400 m³ and an air exchange of 0.5 ACH is to be built. The school is for 500 students (kids), 50 staff, 300 40-W fluorescent lights, 75 computers, 3 printers, and 2 coffee makers. What is the expected sensible-heat gain? What is the expected latent-heat gain?

Problem 8.4

An economizer, shown in Fig. 8.14, is a "free-cooling" system. Commercial buildings often have large heat gains because of lighting, computers, occupants, etc. Therefore, even in cool weather, cooling may be required. Under these conditions, outdoor air may be used for cooling via economizers. Find the equation relating the amount of outdoor air needed as a function of the outdoor air temperature.

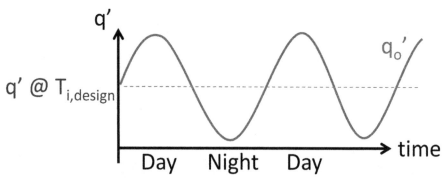

Figure 8.13. The instantaneous heat flux on the outside wall of a building with respect to time (created by D. Ting).

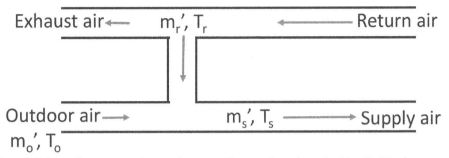

Figure 8.14. An economizer, a free-cooling system (created by D. Ting).

Deduce the equation for the outside temperature above which the economizer is inadequate for cooling.

Problem 8.5

The ventilation air requirement of a building is 4700 L/s, representing 18% of the total supply airflow to the HVAC system. The building supply air temperature is 13°C and the interior temperature of the building is 23°C. What is the minimum outdoor temperature at which an economizer can be used without heating the outdoor air?

Problem 8.6

A two-storey building with a flat roof can be approximated as a rectangular box of 7 m by 12 m by 5 m. The roof is 0.25-m thick and the walls are 0.15-m thick. To ease the heat transmission calculations, they can be assumed to made of fiberglass insulation with $k = 0.06$ W/(m·K). The windows are double-glazed with $U_{glaz} = 3.0$ W/(m²·K) and they cover 40% of the walls. The air exchange rate is 0.5 per hour. Find the heat transmission coefficient, K_{tot}.

REFERENCES

ASHRAE, *2017 ASHRAE Handbook: Fundamentals,* SI Edition., 2017.

J.F. Kreider, P.S. Curtiss, A. Rabl, *Heating and Cooling of Buildings: Design for Efficiency,* 2nd ed., McGraw-Hill, New York, 2002.

T.H. Kuehn, J.W. Ramsey, J.L. Threlkeld, *Thermal Environmental Engineering,* 3rd ed., Prentice-Hall, Upper Saddle River, 1998.

CHAPTER 9

Energy Estimation

"The energy of the mind is the essence of life."

–Aristotle

Nomenclature

ACH	Number of air exchanges per hour
C_d	The degradation coefficient
c_p	Heat capacity at constant pressure
D	Number of degree-days; D_c is the degree-days of cooling, D_h is the degree-days of heating, D_z is the number of degree-days, $D_{z,s}$ is the number of degree-seconds
E	Energy; $E_{required}$ is the required energy
F	Fuel; $F_{required}$ is the required amount of fuel
HV	Heating value
HVAC	Heating, ventilation, and air conditioning

(Continued)

I   Solar radiation; I_{dif} is the diffused solar radiation, I_{dir} is the direct solar irradiance, I_{tot} is total solar radiation

K   Heat transmission coefficient; K_{tot} is the total heat-loss coefficient

k   Heat transfer coefficient; k_{cond} is the total conductive heat transmission coefficient

LF   Load factor

N   Number; N_{bin} is number of days based on the Bin method

PLF   Part load factor

Q   Heat; Q_c is the cooling energy, Q_h is the heating energy, $Q_{h,bin}$ is the heating energy based on the Bin method

Q'   Heat transfer rate; Q_{gains}' is the rate of total heat gain, Q_h' is the heating rate, $Q_{heating,design}'$ is the required heating rate at the design condition, Q_{loss}' is the heat loss rate, $Q_{loss,design}'$ is the heat loss rate at the design condition, Q_{net}' is the net heating rate required

T   Temperature; T_{bal} is the balance-point temperature, T_i is the indoor temperature, $T_{i,design}$ is the design indoor temperature, T_{max} is the maximum balance-point temperature, T_o is the outdoor temperature, $T_{o,avg}$ is the average of T_o, $T_{o,design}$ is the design outdoor temperature, T_Z is the zero-load temperature

t   Time

U   Thermal conductivity, U value; $(UA)_{eff}$ is the total building heat-loss coefficient

Greek and Other Symbols

η Efficiency; η_c is the efficiency of the cooling system, η_h is the efficiency of the heating system

ρ Density

$\forall$ Volume; $\forall'$ is the volume flow rate

9.1 THE NEED TO ESTIMATE ENERGY USAGE

A building's energy requirement, in addition to the design heating and cooling loads, needs to be reasonably estimated because of the following reasons.

1) The building energy usage is required to be in compliance with codes and standards. This is more so these days than ever before as we strive to ensure sustainable buildings and environment into the future.

2) The building energy usage is a critical part of an economic analysis. The annual energy usage is a priori for the minimization of the total heating, ventilation, air conditioning (HVAC) cost. The total cost can be in the form of the total life cycle cost, providing a solid platform for design optimization.

3) An accurate energy usage estimate allows proper comparison of alternate HVAC designs. Although status quo of employing the typical system is an easy routine for an established HVAC company to execute, the progress of Homo sapiens leans on our striving to uniquely improve.

After all, energy means money. Truthfully, a joule of energy saved is much more than a penny saved, as the ripple expands to our environment. Thermodynamically or thermo-economically speaking, saving energy entails the lessening of entropy generation and/or reducing the exergy destruction.

Interestingly, energy savings do not necessarily imply better economics when it comes to commercial and/or industrial buildings. In other words, attempting to save some operation money by lowering the level of thermal comfort can reduce the output of the work, costing substantial revenue loss. On the other hand, investing money to ensure the workers are in the most comfortable condition can lead to substantially more and better output, and thus, revenue. This is also a fact when it comes to schools. The performance of the pupils decreases as the classroom becomes less comfortable and increases as the room is better conditioned. In short, sometimes it is better to invest the energy rather than trying to save it. Or, to quote Constance Chuks Friday, "Energy is precious, use your energy to build not to destroy."

9.2 MODELING ENERGY USAGE

Chapter 19 of the ASHRAE 2017 Handbook [ASHRAE 2017] discloses estimating energy use for two purposes. The first is the classical Forward Modeling, which models buildings and HVAC systems for design optimization. The second purpose is called Data-Driven Modeling. Existing building energy use is modeled for establishing references, deducing retrofit savings, and putting into action model-predictive control.

9.2.1 Forward Modeling

The classical forward approach predicts output with known structure and parameters when the system is subject to specified input variables. The main advantage is that the system does not need to be physically completed. In other words, the model can be utilized to predict the performance of various systems. The results from these performances are used to assist in selecting a better system. Therefore, this approach is best suited in the preliminary design and analysis stage.

Figure 9.1 is a schematic of a generic mathematical model, which aims at describing the system behavior. The thermal mass, mechanical properties of the various elements of the building, etc., describe the system. For example, if the input variable thermostat setting is reduced by 1°C, the model will spit out the amount of energy saved over a particular period of time, such as a mid-winter night. It is interesting to note that climate, not weather, is the uncontrollable variable of concern. This is because we are interested in general annual operation, and not the performance under peculiar weather condition.

To initiate the modeling, the building or component of interest is described physically. For the typical case where the entire building is of concern, its geographical location, dimensions, orientation, building envelop materials and dimensions, the type of HVAC equipment, etc. are detailed.

The model should be based on sound engineering principles. The more recognizable simulation codes include TRNSYS, DOE-2, EnergyPlus, and ESP-r.

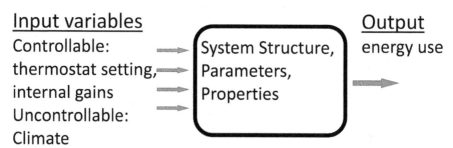

Figure 9.1.   A schematic of the classical forward modeling of a building (created by D. Ting). Changes in weather, thermostat setting, and internal heat gains have a direct impact on the energy usage.

Forward models have the following three main parts:

1) *Space Load.* The sensible and latent heating and cooling demands of the space.
2) *Secondary Equipment Load.* Secondary equipment is the equipment that deliver the heating, cooling, or ventilating medium to the conditioned spaces.
3) *Primary Equipment Energy Requirement.* Primary equipment is the central plant equipment that converts fuel or electric energy to heating and cooling effect.

The first step in forward modeling is to deduce the space load. In this step, the amounts of sensible and latent heat, which need to be removed from or added to the concerned space to ensure thermal comfort of the occupants, are to be deduced. In its simplest form, the space load is only a function of the outdoor dry-bulb temperature. The outdoor humidity ratio, solar heat, internal gains, heat and moisture storage in walls and interior, wind effects on building envelope, heat transfer rate, and infiltration/exfiltration are also included to improve the model.

The second step is to translate the estimated space load into load on the secondary equipment. The base case is to simply assess duct and/or piping losses or gains as a function of the outdoor dry-bulb temperature. The more comprehensive models also provide hour-by-hour simulation of an air system such as variable-air-volume with outdoor-air cooling. All required energy, which encompasses electrical energy to operate fans and/or pumps, energy in heated and chilled water, etc., is included in the more versatile models.

Upon accomplishing the first two steps, the energy required by the primary equipment to meet loads and peak demand on the utility system is determined in the third step. The efficiencies of the equipment and their part-load characteristics are needed. Various forms of energy, electricity, natural gas, oil, etc., are monitored. Relevant codes and standards may be required to convert these energies into source energy or resources consumed. ASHRAE Standard 90.10 is often employed for the economic analysis. Capital costs are included to deduce the life cycle costs, in order to select the more appropriate options.

9.2.2 Data-Driven Modeling

Data-driven modeling is considered the inverse approach. Inverse or reverse engineering is used in the sense that the system behavior is based on the responses, or outputs, with respect to variation in the inputs, see Figure 9.2. The purpose of

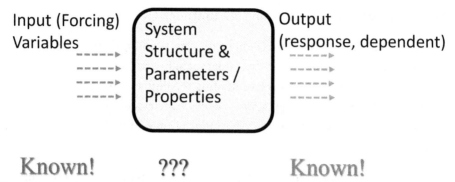

Figure 9.2. A schematic of data-driven modeling (created by D. Ting). Energy usage is a complex function of the involved parameters such as outdoor and indoor air temperature and humidity. Variation in these parameters alter the energy usage via the complex mathematical function.

data-driven modeling is to create a utilizable mathematical description of the system and to estimate system parameters. To do so, the system must already exist. Also, performance data of the system must be available. There are two types of performance data. The first type is the nonintrusive data. These data are gathered under normal operation. To extend the coverage, experiments can be enforced outside the normal operating conditions. In this way, a significantly wider operation range is covered. The more extensive intrusive data allow the engineer to account for expected abnormal operation and to better design the optimal system.

Practically, only limited variables, such as the overall building heat loss coefficient and the time constants associated with the building thermal mass, are measured. Therefore, the number of involved parameters in data-driven models is comparatively smaller than in forward modeling. Nevertheless, these real, in situ data normally allow more accurate capturing of the as-built system performance. Accordingly, data-driven models can more accurately predict future system behaviors under specific circumstances. Solid skills and expertise are required for proper tailoring of data for specific circumstances. In general, data-driven models are less flexible than forward model for estimating energy consumption of different design and operation options.

9.3 ENERGY ESTIMATION METHODS

Simulation software packages have drastically eased HVAC engineers' lives. At the same time, we cannot deny that there is much truth in what Mitch Ratcliffe uttered,

"A computer lets you make more mistakes faster than any other invention in human history, with the possible exceptions of handguns and tequila." Let us stay on the positive side and side with Louis V. Gerstner, Jr. that "Computers are magnificent tools for the realization of our dreams, but no machine can replace the human spark of spirit, compassion, love, and understanding." There is no doubt that some of the time-tested, simple energy-estimation methods deserve continual appreciation in their own right. Furthermore, these "more tangible" methods also aid our understanding of the complex "black-box magic" behind sophisticated software. With that, we will look at the degree-day method followed by the Bin method [Kreider & Rabl, 1994; Kuehn et al., 1998; McQuiston et al., 2005; Mitchell & Braun, 2013].

9.3.1 Degree-Day Method

The degree-day method is simple and yet potent. It assumes that the internal gains and other heat gains such as solar do not vary appreciably and that thermal-storage effects are negligible. Some of the main limitations of this method are the following:

- A single indoor-temperature set point is assumed, i.e., the temperature does not vary with respect to the time of day.
- A constant internal heat-gain rate is assumed. This implies that there is no change with occupancy level or other factors such as turning on or off of appliances and lighting.
- No outdoor-humidity information is provided to estimate latent loads. Strictly speaking, only sensible heat is considered.

With the above backdrop, the building or occupied space heat loss rate decreases linearly with increasing outdoor temperature, see Figure 9.3. The building total heat loss rate,

$$Q_{loss}' = (UA)_{eff} (T_i - T_o),\qquad(9.1)$$

where the **total building heat-loss coefficient**,

$$(UA)_{eff} = Q_{loss,design}' / \left(T_{i,design} - T_{o,design}\right).\qquad(9.2)$$

The heat loss rate at the design condition, $Q_{loss,design}'$, corresponds to the circumstance that the indoor air is at the indoor design temperature[1], $T_{i,design}$, and the outdoor is at the outdoor design condition, $T_{o,design}$. What is delineated in Figure 9.3 is primarily a winter situation, where $T_{o,design}$ is at the start of the x-axis, notably below the indoor temperature, T_i. When the outdoor temperature is at $T_{o,design}$, the total heat loss rate is $Q_{loss,design}'$, the y-intercept of the Q_{loss}' versus T_o line. This minus the total heat gain rate, Q_{gains}', gives the net positive heat loss rate. In short, the required heating rate at the design condition,

$$Q_{heating,design}' = Q_{loss,design}' - Q_{gains}'. \tag{9.3}$$

In general, we can see in Figure 9.3, that the net heating rate required from the heating system is

$$Q_{net}' = (UA)_{eff}(T_z - T_o), \tag{9.4}$$

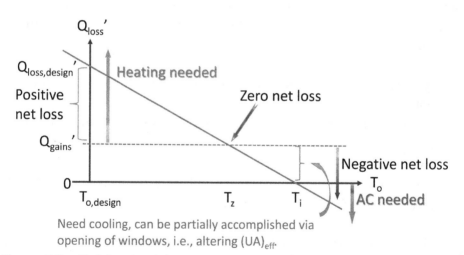

Figure 9.3. Building heat loss rate versus outdoor temperature (created by D. Ting). T_z is the outdoor temperature at which the heat loss from the building is balanced by the internal heat gains.

---

[1]In reality, the indoor design temperature changes with the season. In its simplest form, the clothing we wear during the winter time is much warmer than that in the summer time. For this reason, the indoor design temperature is lower during the cold season, and higher during the warm summer. There are, however, offices where the thermostat is set at the same low temperature during the summer, forcing occupants to put on sweaters. This practice, unfortunately, is not energy sensitive.

where the **zero-load temperature**,

$$T_z = T_i - Q_{gains}{}'/(UA)_{eff}. \tag{9.5}$$

This zero-load temperature is the outdoor temperature that corresponds to the situation when the heat loss rate equals the heat gain rate. With the outdoor temperature at T_z, which is below that of the indoor, T_i, thermal energy transmits from indoor to outdoor. The rate of this heat loss, however, is balanced by the total heat gain rate, $Q_{gains}{}'$. In other words, all needed heating is supplied by the natural gains, and hence, the net heating rate supplied by the HVAC heating system, $Q_{net}{}'$, described by Eq. 9.4, is zero. All of us who have lived in the temperate zone experience this during the transitional seasons. Typically this zero-load temperature is around 18°C, when indoor air stays at the "comfortable age" of 21, i.e., 21°C, with the heating system off.

Continuing with the winter heating case, the amount of heating energy required over a specified time period,

$$E_{required} = (UA)_{eff} D_{z,s}, \tag{9.6}$$

or, where 'seconds' in $D_{z,s}$ is converted into 'days' in D_z,

$$E_{required} = 24 \, (3600) \, (UA)_{eff} D_z, \tag{9.7}$$

where D_z is the number of degree days where the outdoor temperature, T_o, is below T_z. A bountiful amount of data on degree day can be accessed from the ASHRAE Weather Data [ASHRAE, 2017]. With natural gas furnaces dominating the heating systems, at least in North America, the required fuel to supply the heating demand is a piece of necessary information for energy calculations. The fuel requirement,

$$F_{required} = \frac{24 \, (3600) \, (UA)_{eff} D_z}{\eta_{heater} HV} = \frac{24 \, (3600) \, \dot{Q}_{loss,design} D_z}{\left(T_{i,design} - T_{o,design}\right) \eta_{heater} HV} \tag{9.8}$$

where HV is the heating value of the fuel.

EXAMPLE 9.1. ANNUAL NATURAL GAS NEEDED FOR KEEPING A BUILDING WARM

Given: A building in Windsor, Ontario, Canada, is kept warm over the cold winter via a natural gas furnace. The indoor design temperature, $T_{i,design}$,

(Continued)

is 20°C, and the outdoor design temperature, $T_{o,design}$, is -10°C. The corresponding winter design heat loss, $Q_{loss,design}$, is 30 kW. The internal gains add up to 3 kW. The double-stage furnace has an efficiency of 95%. The heating value of the natural gas averages around 48 MJ/kg.

Find: The amount of natural gas needed.

Solution:

The amount of natural gas required can be estimated using Eq. 9.8,

$$F_{required} = 24\,(3600)\,(UA)_{eff}\,D_z/\,(\eta_h HV).$$

From Eq. 9.2, the total heat loss coefficient,

$$(UA)_{eff} = Q_{loss,design}'/\left(T_{i,design} - T_{o,design}\right).$$
$$= 30{,}000/\,(20 - (-10)$$
$$= 1000\ \text{W/°C}$$

The number of degree days is a function of the zero-load temperature, see, e.g., ASHRAE Weather Data [ASHRAE, 2017]. It increases rapidly with decreasing zero-load temperature. Because of this, lowering the thermostat setting can lead to significant savings in heating. The zero-load temperature can be deduced from Eq. 9.5,

$$T_z = T_i - Q_{gains}'/\,(UA)_{eff}.$$
$$= 20 - 3000/1000$$
$$= 17\text{°C}.$$

Let us assume that the number of degree days for $T_z = 17$°C is 2700°C days. With this, we have

$$F_{required} = 24\,(3600)\,(UA)_{eff}\,D_z/\,(\eta_h HV).$$
$$= 24\,(3600)\,(1000)\,(2700)/\,[(0.95)\,(48{,}000{,}000)]$$
$$= 5116\ \text{kg}$$

We see that 5116 kg of natural gas is needed for the entire winter. It is difficult to justify supplying this heating need using electricity via electric heaters. Thermodynamically speaking, electricity is presumably the highest quality energy. Therefore, using it to do the lowest quality work, heating, is a complete waste of its talent.

9.3.2 Bin Method

The Bin method is an improvement from the degree-day method. It divides the outdoor temperature, T_o, data into various temperature increments, or bins, and further subdivides these data into various times of the day. Extensive amounts of data have been compiled and tabulated by ASHRAE; ASHRAE Weather Data Viewer, page 19.7 of the ASHRAE 2017 Handbook [ASHRAE, 2017]. Note that the mean coincident wet-bulb temperature data are also given for latent-load estimation.

Constant Internal Gains Figure 9.4 is a plot of heating rate versus the outdoor temperature. As in Fig 9.3, the rate of heat loss decreases with increasing outdoor temperature. When the outdoor equals the design outdoor temperature, $T_{o, design}$, the heat loss rate is $Q_{loss, design}'$. Subtracting the corresponding internal heat gain rate, Q_{gains}', from the heat loss at design conditions yields the required net heating rate. In practice, the rated heating system capacity is somewhat higher than the design requirement. This is a safety feature for providing enough heat when T_o is somewhat lower than $T_{o, design}$. Also, note that no heating is required as long as the total heat loss is less than the internal heat gains. i.e., $T_0 \geq T_Z$.

Suppose the free thermal energy outdoors is tapped into via an air-to-air heat pump. The use of this green energy system is gaining ground in the mid-latitude United States and other places with moderate winters. The rate of heating supplied by this heat pump is depicted as a dashed line in Fig 9.4. Let us consider the typical case where the outdoor temperature, T_o, is higher than the design outdoor temperature, $T_{o, design}$, but lower than the zero-load temperature, T_z. We see that the heat

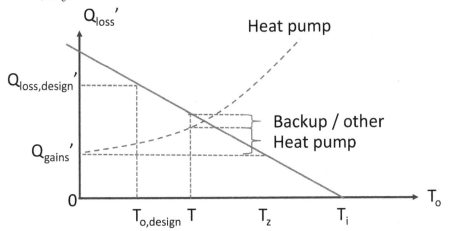

Figure 9.4. Heating rate as a function of outdoor temperature (created by D. Ting).

pump does not supply quite enough needed heating. A backup or other heating system is thus employed to cover the shortfall. The required backup heating is reduced as the weather warms up. On the other hand, the efficiency of the outdoor-air-to-indoor-air heat pump diminishes as the outdoor temperature drops. This is the reason why air-to-air heat pump is not popular in "frozen" places such as Canada.

Most often, heating and air conditioning systems run on partial loads. This is particularly the case in larger buildings. In residential applications, these systems typically operate in the transient on-and-off mode. Let us define a **Load Factor**,

$$LF \equiv Q_{loss}'/Q_{output}'. \tag{9.9}$$

Accounting for equipment-cycling losses, we use the **Part Load Factor**,

$$PLF = 1 - C_d(1 - LF), \tag{9.10}$$

where C_d is the degradation coefficient with a default value of 0.25.

Variable Internal Gains The constant internal gains assumption can be loosened by determining the number of hours in each hour group that are occupied and unoccupied, and/or major appliances, lighting, etc., are on or off. We can introduce something more-or-less identical to the zero-load temperature. We call it the **balance-point temperature**, T_{bal}, which is the value of the outdoor temperature at which the total heat loss is equal to the free heat gain, i.e.,

$$K_{tot}(T_i - T_{bal}) = Q_{gain}', \tag{9.11}$$

where the **total heat-loss (transmission) coefficient**, as introduced in Chapter 8,

$$K_{tot} = k_{cond} + \forall'\rho\, c_p. \tag{9.12}$$

Recall that k_{cond} is the total conductive heat transmission coefficient. The balance-point temperature is thus,

$$T_{bal} = T_i - Q_{gain}'/K_{tot}. \tag{9.13}$$

When the outdoor temperature drops below the balance-point temperature, heating becomes necessary. The required heating rate,

$$Q_h' = (K_{tot}/\eta_h)[T_{bal} - T_o(t)], \tag{9.14}$$

when $T_o < T_{bal}$, and η_h is the efficiency of the heating system. From this, we can calculate the annual energy consumption for heating,

$$Q_h = (K_{tot}/\eta_h) \int [T_{bal} - T_o(t)]_+ \, dt, \tag{9.15}$$

where the + sign indicates that only positive values are to be counted. In other words, when the outdoor temperature is equal to or above, T_{bal}, no heating is needed.

If daily average values of T_o are used, the degree-days for heating (K·°days) is

$$D_h(T_{bal}) = 1 \, day \times \sum_{days} (T_{bal} - T_o)_+ . \tag{9.16}$$

Alternatively, we can use shorter time intervals to estimate the degree-days for heating, i.e.,

$$D_h(T_{bal}) = \frac{1 \, day}{24 \, h} \times 1 \, h \times \sum_{hours} (T_{bal} - T_o)_+ . \tag{9.17}$$

It is advisable to use short time intervals if the time constant (review Transient Thermal Network in Chapter 8) of the building is closer to an hour than to a day.

With the above, the annual heating consumption in terms of degree-days can thus be deduced,

$$Q_h = (K_{tot}/\eta_h) D_h(T_{bal}). \tag{9.18}$$

Degree days or degree hours for a balance-point temperature of 18°C in Europe or 65°F in the United States are widely tabulated. Figure 9.5 plots an arbitrary outdoor temperature as a function of time. The indoor temperature, T_i, is fixed. This, along with constant internal gains, leads to a steady balance-point temperature, T_b. The shaded area under this balance-point temperature and above the outdoor temperature, T_o, is the degree days of heating required,

$$Q_{h,bin} = N_{bin} (K_{tot}/\eta_h) (T_{bal} - T_o)_+ . \tag{9.19}$$

EXAMPLE 9.2. ANNUAL HEATING ENERGY CONSUMPTION

Given: A house in New York with $K_{tot} = 200$ W/K, $Q_{gain}' = 600$ W, $T_i = 21°C$, and $\eta_h = 0.85$.
Find: The required annual heating.

(*Continued*)

Solution:

From Eq. 9.13, the balance-point temperature,

$$T_{bal} = T_i - Q_{gain}'/K_{tot} = 21°C - 600/200 = 18°C.$$

This is the standard T_{bal} value, and hence, the values for many cities, including New York, are tabulated; see, e.g., ASHRAE. For New York with $T_{bal} = 18°C$, D_h = 2800 K days.

The annual heating can be calculated using Eq. 9.18,

$$Q_h = (K_{tot}/\eta_h)\, D_h\, (T_{bal}) = 48\, GJ.$$

It is amazing to see that the most populated city of the United State is so cold. New Yorkers are tough.

Cooling Energy Consumption Analogously, the cooling degree days can be expressed as

$$D_c\, (T_{bal}) = 1\ day \times \sum_{days} (T_{bal} - T_o). \tag{9.20}$$

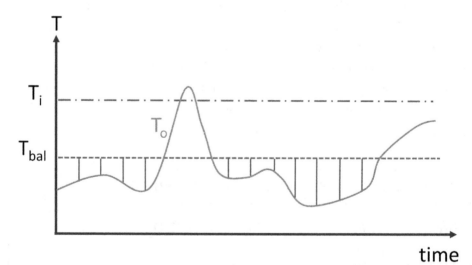

Figure 9.5. Degree-time, or, degree-days, from outdoor temperature versus time plot (created by D. Ting). Heating is needed whenever the outdoor temperature is below the balance temperature.

Here, the negative subscript signifies that cooling is required when the outdoor temperature rises above the balance-point temperature. If cooling and heating degree days are based on the same T_{bal}, then

$$D_h\left(T_{bal}\right) - D_c\left(T_{bal}\right) = 365 \text{ days} \times \left(T_{bal} - T_{o,avg}\right), \qquad (9.21)$$

where $T_{o,\,avg}$ is the annual average of T_o. As such, the annual cooling energy consumption may be estimated analogously,

$$Q_c = \left(K_{tot}/\eta_c\right) D_c\left(T_{bal}\right). \qquad (9.22)$$

In reality, however, it is more challenging to estimate the cooling energy consumption than heating energy consumption, because the total heat transmission coefficient, K_{tot}, value changes. For example, the opening of windows and/or increasing the ventilation during warm days can drastically increase K_{tot}. As this is practically the case, the air conditioning is needed only when the outdoor temperature, T_o, is greater than a temperature corresponding to the balance-point temperature when the total heat transmission coefficient is at its peak; see Fig 9.6. This maximum balance-point temperature,

$$T_{max} = T_i - Q_{gain}'/K_{max}. \qquad (9.23)$$

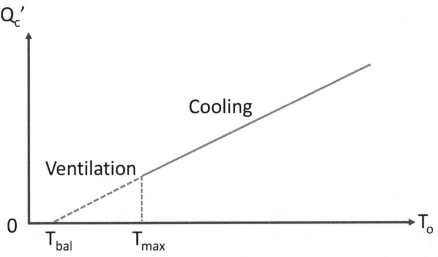

Figure 9.6. Cooling rate as a function of outdoor temperature (created by D. Ting). When the outdoor temperature rises above T_{max}, mechanical cooling by an air conditioning system is required.

Note that K_{tot} for closed windows and/or normal winter ventilation is replaced by K_{max} for opened windows and/or increased ventilation.

To more clearly differentiate the effect of the change in the balance-point temperature on the heating and cooling rate requirement, the heat loss rate versus outdoor temperature is re-plotted in Fig 9.7. It is clear that for a fixed balance-point temperature of T_b, the needed cooling rate, Q_c', is larger than that for a higher summer time balance temperature, T_{max}. It is worth mentioning that the lightening of clothing from winter to summer also contributes to the increase in the balance-point temperature from winter to summer, and hence, saving significant cooling energy.

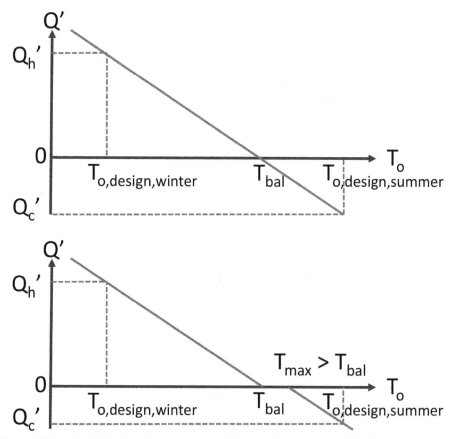

Figure 9.7. Heat loss rate versus outdoor temperature with (a) with a fixed balance-point temperature, T_b, year round and (b) T_b for the heating season and T_{max} for the cooling season (created by D. Ting).

EXAMPLE 9.3. ANNUAL COOLING ENERGY CONSUMPTION

Given: A house in New York with K_{tot} = 200 W/K with closed windows at 0.5 air change per hour, k_{cond} = 145 W/K, Q_{gain}' = 1100 W, T_i = 24°C, and 5 air exchange per hour with open windows.

Find: The required annual cooling.

Solution:

For New York, the degree-days heating,

$$D_h \, (T_{bal} = 18°C) = 2800 \, K \cdot days.$$

and the degree-days cooling,

$$D_c(T_{bal} = 180°C) = D_h(T_{bal} = 180°C) - 365 \, days \, (18°C - 12.5°C)$$
$$= 756 \, K \cdot days.$$

For closed windows, the total heat transmission coefficient,

$$K_{tot} = k_{cond} + \forall' \rho \, c_p = 145 + 0.5 \, ACH$$
$$200 = 100 + 0.5 \, ACH$$
$$1 \, ACH = 200 \, W/K.$$

Therefore,

$$5 \, ACH = 1000 \, W/K.$$

With the windows opened, from Eq. 9.23,

$$K_{max} = k_{cond} + \forall_{max}' \rho \, c_p = 145 + 10 \, ACH$$
$$= 100 + 1000 = 1100.$$

The increased balance-point temperature, as per Eq. 9.23,

$$T_{max} = T_i - Q_{gain}'/K_{max}$$
$$= 24 - 1100/1100$$
$$= 23°C.$$

For New York, the degree-days heating,

$$D_h \, (T_{bal} = 21°C) = 4100 \, K \cdot days,$$

(Continued)

and the degree-days cooling,

$$D_c(T_{bal} = 23°C) = D_h(T_{bal} = 23°C) - 365 \text{ days}(23°C - 12.5°C)$$
$$= 231 \text{ K} \cdot \text{days}.$$

This is approximately 70% less than that without enhanced ventilation of 0.5 ACH!

PROBLEMS

Problem 9.1

Compare the natural gas usage for heating a building located in (a) Toronto, (b) Saint Peterburg, (c) Tianjin, and (d) Copenhagen. Assume that they have the same $(UA)_{eff}$ of 500 W/K, internal gains of 1000 W, indoor temperature set at 21°C, and the furnace has an efficiency of 90%.

Problem 9.2

The indoor air temperature, T_i, is kept at 22°C, and the total internal gain is constant, resulting in a zero load temperature, $T_z = 16°C$. The total building heat loss coefficient, $(UA)_{eff} = 5000$ W/°C. The graph in Fig 9.8 shows the variation of the outdoor temperature over a typical year.

Part I) The required annual heating is_____°C · days.

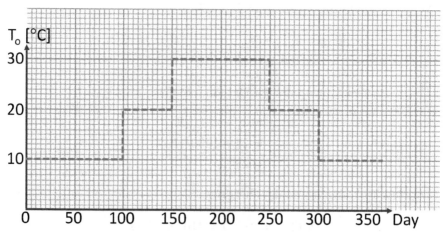

Figure 9.8. Outdoor temperature variation over the year (created by D. Ting).

Part II) The required annual cooling is_____°C · days, assuming constant (UA)$_{eff}$.

Part III) The total internal heat gain rate is_____W.

Part IV) If the total internal heat gain rate is 9000 W instead (with $T_i = 22°C$ and $T_z = 16°C$), then the total heat transmission coefficient K_{tot} of the building would be _____ (include the appropriate units).

The following are review questions. Knowledge learned from previous chapters is required for deducing the solutions.

Problem 9.3

As an HVAC engineer-in-training, your first task is to assist in the design of a greenhouse near Payne Lake (longitude = 75°W, latitude = 60°N), Quebec, Canada. The 3-m tall, 500-m wide and 2000-m deep greenhouse is made of 10-cm thick plexiglass with a thermal conductivity of 1 W/(m K). For simplicity, you may neglect the effects of the framing network and heat loss/gain to/from the ground.

The greenhouse is to be ideally at 25°C and 60% relative humidity. Two air changes per hour is required, and this is provided by infiltration/exfiltration. Five thousand 50-W lights are to be used during the nights and during the days when there is a lack of sunlight. The greenhouse operates around the clock with three 8-hour shifts of workers. Each shift consists of 80 workers; each worker generates 90 W of sensible heat and 60 W of latent heat. The vegetation generates 0.02 kg of moisture into the indoor air per hour per m² of land.

The design condition for the winter is −35°C and 30% relative humidity. For the summer, it is 20°C and 50% relative humidity. The total solar radiation passing through the plexiglass onto the ground inside the greenhouse during the hottest summer hour is $I_{total} = 750$ W/m².

If the indoor air humidity is not a concern, i.e., only the indoor air temperature needs to be maintained at 25°C, what should the capacity of the heating system be? What should the capacity of the cooling system be?

Problem 9.4

An architect has proposed the following greenhouse to a business man in a small town with a longitude of 105°W and a latitude of 66.5°N in the North West Territories of Canada. The greenhouse is a 500-m diameter hemisphere made of 10-cm thick plexiglass with a thermal conductivity of 1 W/(m K), an absorptivity of 0.03, and a reflectivity of 0.07. For simplicity, the effects of the framing network and heat loss/gain to/from the ground can be neglected.

Ideally, the greenhouse is to be at 25°C and 70% relative humidity. One air change per hour is required, and 10% of this is provided by infiltration. Four thousand 50-W lights are to be used during the nights, and during the days when there is a lack of sunlight. The greenhouse operates around the clock with three 8-hour shifts of workers. Each shift consists of 70 workers; each worker generates 90 W of sensible heat and 50 W of latent heat. The vegetation generates 0.02 kg of moisture into the indoor air per hour per m² of land.

The winter design condition is −25°C and 20% relative humidity. The summer design condition is 20°C and 60% relative humidity, with I_{dif} of 200 W/m² and I_{dir} of 500 W/m². To simplify hand calculations, assume that the summer design condition occurs at solar noon of the summer solstice.

Part I) If the indoor humidity is not a concern, i.e., only the temperature needs to be maintained at 25°C, what are the required heating and cooling capacities?

Part II) If both temperature and humidity of the indoors need to be maintained, i.e., 25°C and 70% relative humidity, what are the required heating and cooling capacities?

REFERENCES

ASHRAE, *2017 ASHRAE Handbook,* 6th ed., Fundamentals, Atlanta, *2017* .

J.F. Kreider, A. Rabl, *Heating and Cooling of Buildings: Design for Efficiency,* McGraw-Hill, New York, 1994.

T.H. Kuehn, J.W. Ramsey, J.L. Threlkeld, *Thermal Environmental Engineering,* 3rd ed., Prentice-Hall, Upper Saddle River, 1998.

F.C. McQuiston, J.D. Parker, J.D. Spitler, *Heating, Ventilating and Air Conditioning: Analysis and Design,* 6th ed., Wiley, Hoboken, 2005.

J.W. Mitchell, J.E. Braun, *Principles of Heating, Ventilation, and Air Conditioning in Buildings,* Wiley, Hoboken, 2013.

CHAPTER 10

Heating and Cooling Systems

"The energy of the mind is the essence of life."

–Aristotle

CHAPTER OBJECTIVES

- Recognize forced-air furnaces.
- Understand the workings of a reverse heat engine for heating, ventilation, and air conditioning (HVAC) purposes.
- Recap the thermodynamics of vapor-compression and absorption refrigeration.
- Appreciate the amelioration of HVAC by heat pipes and heat pumps.

Nomenclature

A area

c_p heat capacity at constant pressure

COP   coefficient of performance; COP_{abs} is the coefficient of performance of the absorption refrigeration system, COP_{HP} is the coefficient of performance of the heat pipe, $COP_{HP,Carnot}$ is the Carnot coefficient of performance of the heat pipe, COP_{Ref} is the refrigeration coefficient of performance, $COP_{Ref,Carnot}$ is the Carnot coefficient of performance of the refrigeration

(Continued)

HVAC   heating, ventilation, and air conditioning

m mass; m' is mass flow rate

Q heat; Q_{gen} is the heat for the generator, Q_H is the heat associated with the high-temperature reservoir, Q_{in} is the heat into the system, Q_L is the low-temperature heat, Q_{out} is the heat out of the system

T temperature; T_H is the higher temperature in absolute scale, $T_{i,design}$, is the indoor design temperature, T_L is the lower temperature in absolute scale, $T_{o,design}$ is the outdoor design temperature, ΔT is temperature difference

U thermal conductivity, U-value, $(UA)_{eff}$ is the total building heat-loss coefficient

W work; W_{in} is the work input, W_{out} is the work output, W_{pump} is the work required by the pump

Greek and Other Symbols

η efficiency; η_{Carnot} is the Carnot efficiency, η_{HE} is the heat engine efficiency

η_{II} the second law efficiency; $\eta_{II,HP}$ is the second law efficiency of the heat pipe, $\eta_{II,Ref}$ is the second law efficiency of the refrigeration system

10.1 HEATING VIA FORCED-AIR FURNACES

Thermodynamically speaking, the best way to furnish warmth for thermal comfort is to use the lowest form of energy, i.e., heat. According to the second law of thermodynamic, electricity is the highest quality energy; it can be exploited to do all kinds of work, and thus, should be used for performing high-quality work. At the other end of the energy quality scale is thermal energy or heat. For this reason, it is called the dustbin of all forms of energy. On that account, using electric heaters for heating is just about the worst engineering practice. It is good neither for our pocketbook nor for our environment. It is no wonder combustion still dominating the heating aspect of engineering thermal comfort. Thankfully, more environmentally friendly ways of warming occupied spaces are making significant inroads. Goethermal and/or heat pumps will be briefly discussed in this chapter.

Furnaces, gas furnaces specifically, are still the prevalent system for keeping occupants and occupied spaces warm during the cold winter. Other than ASHRAE, specifically the *2008 ASHRAE Handbook* [2008], this fundamentally well-established system is only lightly covered in HVAC books such as Reddy et al. [2016] and Howell [2017]. The inner workings and maintenance of these heating systems can be found in the more applied technical trade books. Essentially, a gas furnace supplies the warming side of thermal comfort by heating air that is forced through it without direct contact with the scorching fire resulting from the exothermic combustion of the fuel and air mixture.

The simplest form of a gas furnace may be envisioned as a gas fireplace, as portrayed in Fig. 10.1. The combustion process is enclosed within a cabinet. For the fireplace shown, natural gas, along with the required "combustion air"[1], enters the confined combustion chamber. Upon combustion, the flue gas is exhausted via an outlet up through the chimney. The cool room air enters through the slots at the bottom. Without mixing with the combustion gases, this room air picks up heat, via convection-conduction-convection, as it goes around the enclosure

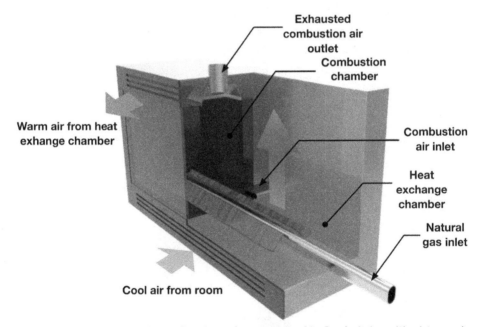

Figure 10.1. A natural gas fireplace (created by N. Cao). It is critical to make sure that the flue gas is exhausted out of the occupied space.

---

[1]The required "combustion air" is generally the stoichiometric amount which contains just enough oxygen to react with the number of fuel molecules. For example, every methane (CH_4) consists of one carbon bonded with four hydrogen atoms. As such, we need two oxygen atoms for the carbon to form one carbon dioxide (CO_2), and two more oxygen to form two water molecules (H_2O).

surrounding the combustion chamber. The rising warm air then exits through the slots at the top of the fireplace.

Forced-air gas furnaces have additional features including a blower that draws the cooler air into the furnace and forces the heated air out into the air distribution duct-work (Fig. 10.2). Compared to the natural gas fireplace depicted in Fig. 10.1, the non-mixing heat exchange is significantly more extensive in the gas furnace. This is more so for today's high-efficiency furnaces, which typically utilize two heat exchangers. The second heat exchanger in a condensing furnace effectively captures the latent heat of vaporization, as the water vapor in the flue gas condenses into liquid water. Furthermore, a variable-speed blower and dual burners are also exploited to mitigate the losses associated with turning the burner and/or the fire on-and-off. Alleviating these transient losses can result in efficiencies way above 90%. It is worth noting that some openings through the building envelope are vital when the furnace is on high. Without any cracks, the flue exhaust cannot readily make its way out due to back pressures, leaving some of the deadly carbon monoxide and other exhaust species inside the occupied space. This is particularly important when other exhaust fans are on high, creating a negative indoor pressure.

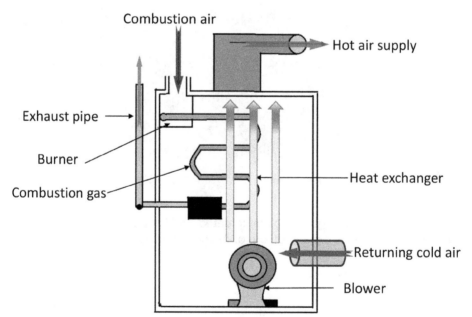

Figure 10.2.   Forced-air gas furnace (created by S.K. Mohanakrishnan, edited by D. Ting).

10.2 REVERSE HEAT ENGINE

A heat engine is an engineering system that converts thermal energy, heat, into mechanical energy for useful work. Figure 10.3 portrays the available thermal energy in terms of a hot reservoir at a high temperature, T_H. The second law of thermodynamics ensures that there is a portion of the energy that cannot be tapped into, and this is the waste heat. This waste heat, labeled as Q_L in Fig. 10.3, has become a hot topic in recent years. Much effort has been invested into recovering a portion of this thermal energy for utility, including transforming it into high-quality electricity via a thermoelectric energy generator. Invoking the first law of thermodynamics gives

$$W_{out} = Q_H - Q_L, \tag{10.1}$$

i.e., the amount of work produced is equal to the quantity of heat input minus that of the amount of waste heat out.

Other than character, we appraise a person based on her or his performance. Performance is the desired output over the required input. As such, the best performing individuals are expected to work hard, as a certain amount of input is required to get an output. Moreover, these high performers must also work smart, producing the maximum desired output with the minimum input, to keep them outstanding. The same is true when it comes to evaluating the performance of engineering

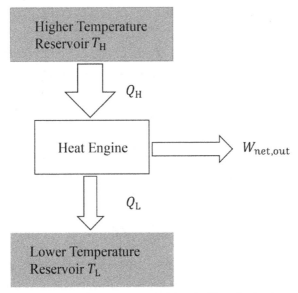

Figure 10.3.   A heat engine (created by Y. Yang). The desired output is work.

systems. For a heat engine, the desired outcome is the work output, W_{out}, and the required input is the heat input from a high-temperature source, Q_H. It follows that the (thermal) efficiency of a heat engine,

$$\eta_{HE} = W_{out}/Q_H. \tag{10.2}$$

From Eq. 10.1, we can rewrite this in terms of waste heat as

$$\eta_{HE} = (Q_H - Q_L)/Q_H. \tag{10.3}$$

As Q_L cannot be zero, according to the second law of thermodynamics, the efficiency of a heat engine must be no more than one.

For engineering thermal comfort, we are more interested in the heat engine operating in reverse. A **reverse heat engine** is a heat engine operating thermodynamically in the reverse direction as a heat engine. Instead of producing work, some work, W_{in}, is inputted into the system so that it can draw heat, Q_L, from the lower temperature space and transfer it to a higher temperature region. To satisfy the first law of thermodynamics, the amount of heat transferred into the higher temperature region,

$$Q_H = Q_L + W_{in}. \tag{10.4}$$

Within the HVAC context, air conditioners, refrigerators, and heat pumps are the most common reverse heat engines for cooling and heating (Fig. 10.4). The name of the system is defined by its function. The function of a heat engine is to produce work, the desirable output. For an air conditioner or a refrigerator, the purpose is to remove heat from the lower temperature space. The objective of a heat pump, on the other hand, is to draw heat from a lower temperature environment.

For a refrigeration system, the desired output is the cooling load, Q_L, and the required input is the work input, W_{in}. In this case, the desired output can, in fact should, be larger than the required input. Therefore, the typical efficiency with an upper limit of one will not suffice, i.e., an efficiency of way over 100% is expected. The coefficient of performance, as defined as

$$COP_{Ref} = Q_L/W_{in} \tag{10.5}$$

is the appropriate performance indicator. Putting the second law of thermodynamics into effect here emanates a finite amount of work input. Namely, W_{in} cannot be zero, and thus, COP has to be less than infinity.

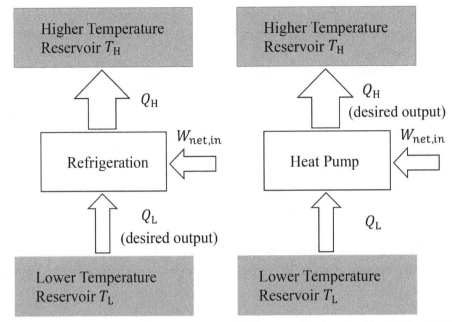

Figure 10.4. Reversing the operation of a heat engine (created by Y. Yang). The desired output can be heat removal or heat addition.

10.3 COOLING FOR THERMAL COMFORT

Refrigeration is the transfer of thermal energy (heat) from a lower temperature space to a higher temperature environment, typically outdoors. According to the second law of thermodynamics, this cannot happen naturally; some external work is required to realize this. The engineering system for realizing this is the refrigerator. In short, a **refrigerator** is a device that removes heat from the conditioned space, and rejects it, along with the thermal energy associated with the needed work input, to the ambient. The **refrigeration cycle** is the thermodynamic cycle in which this is accomplished. Refrigeration and air conditioning are covered in detail in standard thermodynamics textbooks such as Balmer [2011], Borgnakke and Sonntag [2009], and Çengel and Boles [2014]. The larger and/or more sophisticated commercial systems are expounded in the ASHRAE 2008 and 2018 Handbooks [ASHRAE, 2008; ASHRAE, 2018]. Therefore, we will only highlight the basics here. Some authors differentiate air conditioning from refrigeration based on the "customers" the system is serving, i.e., air conditioning serves primarily living human beings, whereas refrigeration keeps, above all, lifeless meat and vegetables fresh. In this book, however, we do not discriminate the customers, may they be breathing creatures, lifeless, or frozen food. We use air conditioning and refrigeration inter-changeably to denote the removing of heat to keep the concerned compartment cool, cold, or frigid.

The three common means for producing refrigeration are:

1. *Mechanical-vapor compression systems.* This is still the most prevailing refrig-eration system. Upon receiving heat from the space to be conditioned, the refrigerant is vaporized and compressed in its vapor phase, before it is con-densed by rejecting heat.
2. *Absorption systems.* The vaporized refrigerant is absorbed, typically into a liq-uid, before it is compressed. Free or low-cost heat such as solar and waste heat, when readily available, make this system particularly attractive.
3. *Gas-compression systems.* These systems are applied in liquefaction, storage, and separation of gases. The refrigerant stays in the gaseous phase throughout the cycle.

Other refrigeration systems worth mentioning include:

1. *Cascade refrigeration.* A cascade of refrigeration cycles is applied to bring the temperature progressively lower from one refrigeration cycle to the next. This can effectively drop the temperature substantially.
2. *Thermoelectric refrigeration.* This is a thermocouple functioning in reverse. With the two junctions of two dissimilar materials at different temperatures, an electro-motive force is produced, enabling the thermocouple to give out a reading corresponding to these temperatures. In reverse, passing an electric current through two dissimilar materials causes the temperature at one junc-tion to drop, whereas that of the other to rise.

10.3.1 Vapor-Compression Refrigeration

The ideal vapor-compression cycle corresponding to the most common refrigera-tion system is illustrated in Fig. 10.5. It consists of

1. Isentropic (constant-entropy) compression
2. Isobaric (constant-pressure) heat rejection
3. Adiabatic (constant-enthalpy) throttling
4. Isobaric (constant-pressure) heat absorption

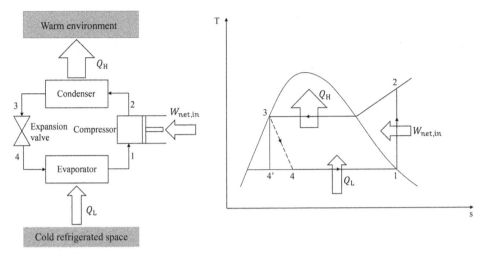

Figure 10.5.   The ideal vapor-compression refrigeration cycle, where the actual deviation is depicted by the dashed lines (created by Y. Yang).

We note that the throttling process is an irreversible, but assumed to be adiabatic, process. Under steady-flow conditions, the energy balance gives

$$Q_{in} - Q_{out} + W_{in} - W_{out} = H_{out} - H_{in}, \tag{10.6}$$

where Q denotes heat, W, work, and H, enthalpy. Normally, there is no work output in a vapor-compression refrigeration system, i.e., W_{out} is zero. Therefore, we have

$$Q_{out} = Q_{in} + W_{in}. \tag{10.7}$$

That is, the amount of heat that needs to be rejected is equal to that drawn from the space to be conditioned plus the energy required for compressing the vapor. This is what Eq. 10.4 says, where Q_H is Q_{out} and Q_L is Q_{in}. Similarly, the coefficient of performance,

$$COP_{Ref} = Q_{in}/W_{in}, \tag{10.8}$$

where the heat into the refrigeration system is the heat removed from the space to be cooled.

   In real life, there are non-idealities such as frictional losses along the passages and heat losses. Therefore, the isentropic processes are, in reality, processes with some increase in entropy (Fig. 10.5); notwithstanding that a hot fluid can lose entropy via heat loss. The ideal constant-pressure flow of working fluid shows up as decreasing-pressure process lines.

10.3.2 Absorption Refrigeration

The main energy usage in vapor-compression refrigeration is to compress the refrigeration in its vapor phase after it picks up heat from the space of concern. From our everyday living, we appreciate that it takes quite a bit of work to compress atmospheric air into a tire. As such, replacing the compressor with something that does not consume as much energy can lead to significant energy savings. This is the case in absorption refrigeration, shown in Fig. 10.6, where the compressor is replaced by an absorber. The absorber is made up of a pump, a heat exchanger, a generator, a valve, and a rectifier. As liquid, instead of vapor, is the working fluid that needs to be moved, it requires considerably less energy to run the pump than the counterpart compressor in the vapor-compression system. Ammonia–water, water–lithium bromide, and water–lithium chloride are the most common working fluids [Aman et al., 2019].

Let us look at the ammonia–water system depicted in Fig. 10.6. After picking up the heat from the space of interest, ammonia vapor moves into the absorber. It reacts with water to form an ammonia–water solution while releasing heat. To promote the reaction, some heat is removed in practice. The liquid ammonia–water

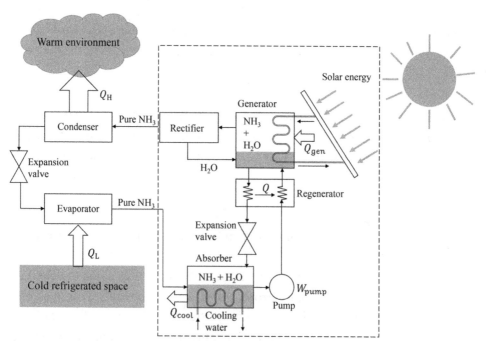

Figure 10.6. Absorption refrigeration (created by Y. Yang).

solution is then pumped to the generator. In the generator, heat is employed to vaporize the solution. A rectifier is used to separate the water from the ammonia-rich vapor. The denser water returns to the generator, whereas the ammonia vapor moves onward, into the condenser, where heat is rejected. With the removal of heat, the high-pressure, high-purity ammonia vapor condenses in the condenser, before it subsequently expands through a throttling valve.

For the ammonia–water absorption refrigeration system, heat is needed to vaporize the solution in the generator. This heat is thus a required input, as far as performance is concerned. Accordingly, the coefficient of performance of an absorption refrigeration system is

$$COP_{abs} = Q_L / \left(Q_{gen} + W_{pump} \right). \tag{10.9}$$

The pump is for moving liquid ammonia, not for compressing a vapor, and hence, the pumping work is very small. Neglecting the pumping work, we may approximate the coefficient of performance as

$$COP_{abs} \approx Q_L / Q_{gen}. \tag{10.10}$$

It is clear that a fair amount of heat, Q_{gen}, is required to operate the system. Therefore, an absorption refrigeration system is most appealing when free or waste heat at 100 to 200°C is available and can be easily capitalized. Incidentally, this is typically the case when air conditioning is most needed, i.e., during the hot summer afternoon when solar heat is abundant. The solar thermal energy can be further tapped into, via a solar bubble pump [Aman et al., 2018], for example, to furnish the needed energy to drive the pump, i.e., W_{pump}. As the efficiency of solar bubble pumps is low, the higher-capital solar photovoltaic may be worth the investment.

10.4 HEAT PIPE VERSUS HEAT PUMP

The most sustainable approach to provide thermal comfort is to realize it by tapping into as much free and renewable energy as possible. This can be accomplished in a straightforward manner via a heat exchanger, when heat needs to be removed from an occupied space when the ambient is at a lower temperature, and also when heat is needed for thermal comfort when the outdoors is warmer. As the space to be conditioned may be at some distance from the outdoors, the conventional heat exchanger is not effective in executing the heat transfer. An ingenious way to overcome this lies in an invention called the heat pipe. A **heat pipe** is a heat exchanger that utilizes high thermal conductivity and phase changes to forcefully transfer heat between two solid interfaces (Fig. 10.7). Capillary action is put into effect via robust wicks that effectively circulate the liquid separated over a distance at two different temperatures. It is clear that a heat pipe performs best when the capillary force is complemented by gravity.

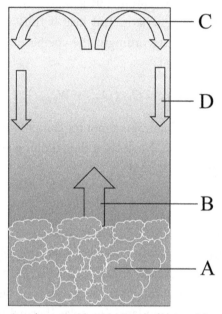

Figure 10.7.   A heat pipe (created by Y. Yang): (A) heat is absorbed in the evaporating section; (B) liquid boils and forms vapor; (C) vapor releases heat and condenses into liquid; and (D) cooled liquid returns, via capillary force and gravity, to the heat absorbing section.

EXAMPLE 10.1. FREE PREHEATING AND REHEATING WITH A HEAT PIPE

Given: As covered in Chapter 4, preheating air can improve humidification, and reheating (after humidification) is a versatile means to control the humidity in the winter time. In a similar token, reheating cold air after passing through a cooling coil is desirable in the hot and humid summer. Unfortunately, whatever heat that is added into the air distribution system ultimately needs to be removed. This can add significant stress on the cooling system. A simple but potent remedy is to exploit a heat pipe as shown in Fig. 10.8. Hot and humid summer air at 1 kg/s is pre-cooled by the heat pipe from 35°C and 80% relative humidity to 30°C before entering the cooling coil. After passing through the cooling coil, the air is at 10°C and 95% relative humidity. It is then raised to 15°C by the free heat extracted from the incoming summer air via the heat pipe.

Find: The rate of cooling energy saved.

Solution:

Detailed calculations can be performed with the help of a pschrometric chart. As a first estimate, the rate of cooling energy saved is

$$Q' = m' c_p \Delta T$$
$$= 1 \, \text{kg/s} \, (1.00 \, \text{kJ/kg} \cdot \text{K}) \, (35°C - 30°C)$$
$$= 5 \, \text{kW}$$

This can be huge!

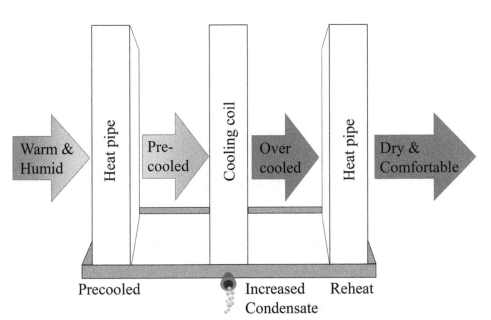

Figure 10.8. A heat pipe for precooling and reheating of hot and humid summer air (created by Y. Yang).

Often, the transport of the working fluid requires boosting beyond which the capillary force can provide. This is the case when the distance between the space of concern and the reservoir, to dump heat into or gather heat from, is large. To realize thermal energy transport enhancement, a pump is added (Fig. 10.9). With a working pump, the intricate wicks are no longer needed; in fact, wicks will slow down the flowing fluid and add to the pumping cost because of significant pressure loss. The invention of this simple and efficacious heat pump also dawned geothermal energy systems. Another everyday example is the air-to-air heat pump; its effective operation leans on heightened heat transportation.

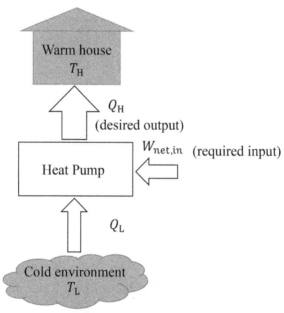

Figure 10.9.   A heat pump (created by Y. Yang).

EXAMPLE 10.2. ELECTRIC HEATER VERSUS HEAT PUMP

Given: A new house owner is contemplating purchasing a heating system for the upcoming winter. His former engineering schoolmate who is in the business of HVAC systems proposes to him a slick electric heater with an efficiency near 100%. To be precis, all the electricity that the heater consumes is ultimately converted into heat. Another former school-mate of his owns a heat pump business. The required heating capacity is 5000 W. A suitable heat pump runs on 20% of the rated capacity, i.e., 1000 W of electricity.

Find: The performance of the two systems.

Solution:

For the electric heater,

$$\eta_{EH} = Q_H/W_{in} = 100\%$$

This is also the coefficient of performance, i.e.,

$$COP_{EH} = Q_H/W_{in} = 1.$$

For the heat pump,

$$COP_{HP} = Q_H/W_{in} = 5000/1000 = 5$$

Evidently, the heat pump performs five times better than the electric heater.

Regarding geothermal energy for heating and cooling of buildings, many advancements have been achieved in recent years, and much is left to be furthered. For the smaller-footprint, vertical system, the working fluid is pumped into a bore-hole, either transferring heat into the ground or capturing heat from the ground, before it returns. Conventionally, a U-tube is encompassed within the borehole to house the flowing fluid. Coaxial tube, Fig. 10.10, has recently been found to be supe-rior in doing the job [Cvetkovski et al., 2014; Gordon et al., 2017]. It is speculated that the larger tube area for heat transfer with the ground, and the better insulation between the outgoing and returning fluid between the inner pipe and the annulus, are some of the reasons for the consequential improvement.

10.5 PUSHING FOR THE BEST PERFORMANCE

Example 10.2 illustrates that 100% is not necessarily the achievable best, i.e., there is room to perform way better. Then, the question is how far can we push for the utmost possible performance? To answer this, we need to know the theoretical best. Only then, can we set that as our goal and strain toward it. Let us start with the heat

Figure 10.10. Geothermal heat pump, the emerging superior co-axial heat exchanger (created by D. Gordon).

engine (Fig. 10.4). The best that a heat engine can accomplish is to suppositionally complete the cycle via four reversible processes. To our knowledge, this ideal heat engine cycle was first proposed by French engineer Sadi Carnot, in 1824. Appropriately, this theoretical cycle is named after him as the Carnot cycle. It is worth noting that the efficiency of a Carnot cycle does not depend on the working fluid; it is solely a function of the reservoir temperatures. With the high-temperature source at T_H and the low-temperature reservoir at T_L, the thermal efficiency of the Carnot cycle,

$$\eta_{Carnot} = 1 - Q_L/Q_H = 1 - T_L/T_H, \tag{10.11}$$

where the temperatures T_L and T_H are absolute temperatures.

Our interest here is the reverse heat engine that has seriously benefitted HVAC engineering. Carnot refrigeration cycle is the reverse of the Carnot heat-engine cycle. As such, the ideal refrigeration cycle has the Carnot refrigeration coefficient of performance:

$$COP_{Ref,Carnot} = Q_L/(Q_H - Q_L) = T_L/(T_H - T_L). \tag{10.12}$$

Recall that for refrigeration, the desired output is Q_L and the required input work is equal to Q_H minus Q_L. A heat pump is used to supply heat from a lower temperature reservoir. By this very nature, the desired output is Q_H, and the required work is equal to Q_H minus Q_L. Therefore, the COP of an ideal heat pump is

$$COP_{HP,Carnot} = Q_H/(Q_H - Q_L) = T_H/(T_H - T_L). \tag{10.13}$$

The real coefficients of performance for a refrigerator and a heat pump are, respectively,

$$COP_{Ref} = Q_L/(Q_H - Q_L) \tag{10.14}$$

and

$$COP_{HP} = Q_H/(Q_H - Q_L) \tag{10.15}$$

The values of these coefficients of performance are less than their corresponding golden standards, i.e., Carnot coefficients of performance. Also, we see that

$$COP_{HP} = COP_{Ref} + 1 \tag{10.16}$$

With the performance indicators properly defined above, we can improve the evaluation of the system performance beyond that inadequately provided by the first law efficiency alone. Such a performance gauge is called the second law efficiency. For a refrigeration system, it is simply

$$\eta_{II,Ref} = COP_{Ref}/COP_{Ref,Carnot}. \tag{10.17}$$

Likewise, the second law efficient of a heat pump is

$$\eta_{II,HP} = COP_{HP}/COP_{HP,Carnot}. \tag{10.18}$$

EXAMPLE 10.3. HEAT PUMP PERFORMANCE AS A FUNCTION OF SPACE-RESERVOIR TEMPERATURE DIFFERENCE

Given: A house is to be kept at 20°C while the outdoors is at 10°C. A 500 W heat pump is able to meet the need by furnishing 2000 W of heat.

Find: The performance of the system.

(Continued)

Solution:

The Carnot coefficient of performance of the heat pump:

$$COP_{HP,Carnot} = T_H/(T_H - T_L)$$
$$= 293.15/(293.15 - 283.15)$$
$$= 29.32$$

The actual coefficient of performance:

$$COP_{HP} = Q_H/(Q_H - Q_L)$$
$$= 2000/500$$
$$= 4$$

Thus, the second law efficiency,

$$\eta_{II} = 4/29.32 = 13.6\%$$

We see that while the heat pump is four times better than its electric heater counterpart, there remains much room for further improvement.

Assume that the same 2000 W of heating to keep a better insulated house at 20°C when the outdoors is at −10°C can be supplied by the same heat pump. What can you say concerning the performance of the heat pump operating at this colder outdoor condition?

The theoretical best for the 10°C colder outdoors is

$$COP_{HP,Carnot} = T_H/(T_H - T_L)$$
$$= 293.15/(293.15 - 263.15)$$
$$= 9.77$$

With COP_{HP} of two, the second law efficiency becomes

$$\eta_{II} = 4/9.77 = 41\%$$

The second law efficiency has tripled! Practically, this tripling may only be achieved with substantial innovation, i.e., a much better, and thus, costly, heat pump is needed. In other words, three heat pumps, instead of just one with the same COP of four, are required when the outdoor temperature drops from 10°C to −10°C. These three heat pumps operate at 1500 W, only 500 W less than the presumably much cheaper electric heater. In short, when the temperature difference increases, heat pumps become less appealing.

PROBLEMS

Problem 10.1

A building is kept warm over the cold winter via a natural gas furnace. The indoor design temperature, $T_{i,design}$, is 20°C, and the outdoor design temperature, $T_{o,design}$, is –10°C. The corresponding winter design heat loss, $Q'_{loss,design}$, is 30 kW. The internal gains add up to 3 kW. The furnace has an efficiency of 95%, and the heating value of the natural gas is 48 MJ/kg. What is the rate of natural gas needed for the furnace? What is the corresponding rate of "combustion air" required?

Problem 10.2

Compare the rate of natural gas usage for heating a building located in (a) Toronto, (b) Saint Peterburg, (c) Tianjin, and (d) Copenhagen on a typical winter night. Assume that they have the same $(UA)_{eff}$ of 500 W/K, internal gains of 1000 W, indoor temperature set at 21°C, and the furnace has an efficiency of 90%.

Problem 10.3

How much heating can be provided by 1 kg of methane? What is the difference in the amount of heating without and with condensing the water vapor into liquid water?

Problem 10.4

A building gains 8 kW of heat in a typical summer day. This heat gain is to be removed by a vapor-compression refrigeration system that operates on refrigerant-134a, where the evaporator is at 350 kPa and the condenser is at 750 kPa. What mass flow rate of the refrigerant is appropriate? What is the required capacity?

Problem 10.5

An air-to-air heat pump operates on the vapor-compression refrigeration cycle, using refrigerant-134a, to keep a space warm using cooler outdoor air. The refrigerant enters the condenser at 45°C and 900 kPa at a rate of 0.02 kg/s. After rejecting heat into the occupied space, it leaves the condenser at 840 kPa and is sub-cooled by 4°C. After picking up the outdoor heat via the evaporator, the refrigerant enters the compressor at 240 kPa and is superheated by 5°C. What is the heating rate? What is the coefficient of performance of the heat pump?

Problem 10.6

Solar heat is supplied to the generator of an absorption refrigeration system at 180°C at 10 kW. The system is used to cool the occupied space to 20°C, while the outdoors is at 35°C. What is the heat removal rate if the coefficient of performance of the absorption refrigeration system is 0.5?

REFERENCES

J. Aman, P. Henshaw, D.S-K. Ting, "Performance characterization of a bubble pump for vapor absorption refrigeration systems," International Journal of Refrigeration, 85: 58–69, 2018.

J. Aman, P. Henshaw, D.S-K. Ting, "Enhanced exergy analysis of a bubble-pump-driven LiCl-H_2O absorption air-conditioning system," International Journal of Exergy, 28(4): 333–354, 2019.

ASHRAE, *2008 ASHRAE Handbook: HVAC Systems and Equipment*, ASHRAE, Atlanta, Georgia 2008.

ASHRAE, *2018 ASHRAE Handbook: Refrigeration*, ASHRAE, Atlanta, Georgia 2018.

R.T. Balmer, *Modern Engineering Thermodynamics*, Academic Press, Burlington, 2011.

C. Borgnakke, R.E. Sonntag, *Fundamentals of Thermodynamics*, 7th ed., Wiley, Hoboken, 2009.

Y.A. Çengel, M.A. Boles, *Thermodynamics: An Engineering Approach*, 8th ed., McGraw-Hill, New York, 2014.

C.G. Cvetkovski, S. Reitsma, T. Bolisetti, D.S-K. Ting, "Ground source heat pumps: should we use single U-bend or coaxial ground exchanger loops?," International Journal of Environmental Studies, 71(6): 828–839, 2014.

D. Gordon, T. Bolisetti, D.S-K. Ting, S. Reitsma, "Short-term fluid temperature variations in either a coaxial or U-tube borehole heat exchanger," Geothermics, 67: 29–39, 2017.

R.H. Howell, *Principles of Heating, Ventilating and Air Conditioning*, 8th ed., ASHRAE, Atlanta, Georgia, 2017.

T.A. Reddy, J.F. Kreider, P.S. Curtiss, A. Rabl, *Heating and Cooling of Buildings: Principles and Practice of Energy Efficient Design*, 3rd ed., CRC Press, Boca Raton, Florida, 2016.

CHAPTER 11

Nature Thermal Comfort

"Look deep into nature, and then you will understand everything better."

−Albert Einstein.

CHAPTER OBJECTIVES

- Appreciate that thermal comfort is imperative for survival and thriving.
- Understand the ingenious intelligent designs in nature for maintaining thermal comfort.
- Adopt nature to design energy-efficient buildings.

Nomenclature

HVAC Heating, ventilation, and air conditioning

P Pressure; P_s is stack pressure, ΔP is pressure difference, ΔP_s is the indoor–outdoor pressure difference due to stack effect

$\forall$ Volume; $\forall'$

11.1 EFFECTIVE THERMAL COMFORTING IN MAMMAL REPRODUCTION

Ever wonder how a fetus avoids heatstroke inside the mother's womb? The skin of a healthy human is at 33°C, whereas the body is around 37°C, see Fig. 11.1. These temperatures are so consistent that since antiquity, they have been used as an index of illness [Clark, 1984]. A deviation of more than ±3.5°C from the normal

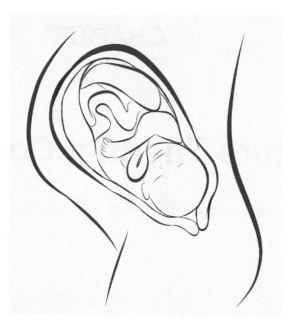

Figure 11.1. How does a fetus avoid heatstroke? (created by S. Akhand).

37°C body temperature can result in physiological impairments and fatality [Lim et al., 2008]. In the subtropics, a healthy individual at these skin and body temperatures is comfortable sleeping naked at a relatively warm temperature around 30°C and 50% relative humidity [Lin & Deng, 2008]. Based on these facts, the mother's womb, at 37°C, is like a sauna. As such, one would think that for the fetus to be thermally comfortable, a substantial amount of heat must be removed via the many blood vessels of the placenta. Although it is true that the amazing placenta is marvelous in supplying all the needs of the baby, including appropriate heat removal, the comfortable temperature for a fetus, however, is quite a bit warmer than that for post-birth beings. Apparently, fresh newborns are thermally comfortable at a temperature that is close to mother's body temperature [Hey & Katz, 1970]. This comfortable temperature rapidly drops after birth so that within a week, it stabilizes to merely a degree or so above that which is comfortable for an average naked sleeping adult. This is but one of the myriad of fascinating ways nature takes care of us. No wonder Max Planck, after devoted his entire life to science, exclaimed, "We must assume behind this force the existence of a conscious and intelligent mind." This sentiment was echoed by great minds such as Sir Fred Hoyle, who uttered, "A common sense interpretation of the facts suggests that a super-intellect has monkeyed with the physics, as well as with chemistry and biology, and that there are no blind forces worth speaking about in nature. The numbers one calculates from the facts seem to me so overwhelming as to put this conclusion almost beyond question."

Along the same line, Rommel et al. [1998] nicely disclosed how male marine mammals ensure their pedigree is secured without a scrotum for keeping their testes cool and healthy. Male cetaceans and seals realize this by very much the same mechanism that keeps the fetus cool in a pregnant female. One could infer from this that tight briefs are not good for men, especially those who wish to father many offspring. For this and other good health reasons, comfy loose boxers have been recommended by some experts.

11.2   WARM-BLOODED VERSUS COLD-BLOODED CREATURES

Warm-blooded creatures are inflexible when it comes to body temperature. Regardless of the temperature of the surroundings, their body temperatures have to stay, within tight margins, at a specific temperature to remain viable and healthy, see Fig. 11.2. Cold-blooded animals, on the other hand, are very adaptable. They let their body temperature equilibrate to that of the environment that they are in, and can survive over a wide range of temperatures.

To adhere to the definitive body temperature, homeotherms (warm-blooded creatures) regulate their metabolic heat output over a wide range, see Fig. 11.3. Creatures such as humans resort to sweating when it is hot, and shivering when it is too cold. As they can tolerate a much wider temperature change, poikilotherms (cold-blooded creatures) by-and-large do not sweat when it is hot. More notably, their body temperatures can adjust down to the freezing point. In fact, fish can survive in slightly sub-zero temperatures. Like other cold-blooded creatures, fish also tap into solar energy. When they are in cold water, sun-basking can raise their temperatures, and this can improve their fitness and swimming [Nordahl et al., 2018].

11.3   HOW NATURE PREVAILS IN DIFFERENT ENVIRONMENTS

Plants in hot and humid rainforests: In rainforests, a far-reaching variety of plants thrive. The forest plants worry not about the essentials for survival. The heavens unceasingly lavish down plenty of sunlight and rainwater day after day. The moisture keeps the temperature of the sun-strong forest in check, furnishing the thermally comfortable conditions for the vegetable kingdom to blossom year round. Thereupon, vegetarians and creatures such as the jungle people (orangutans) and hornbills (the celestial messenger of the upper world according to the Ibans and

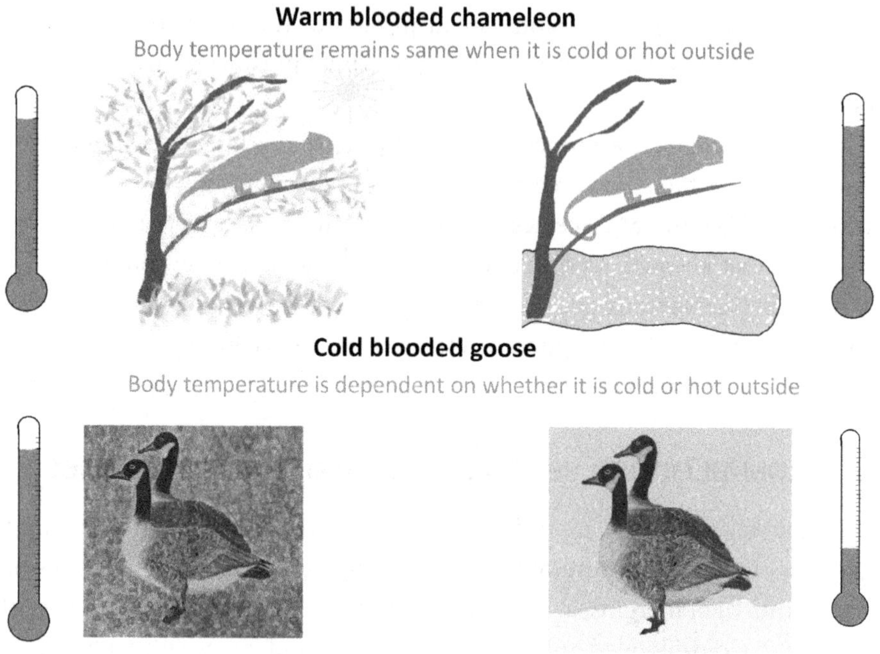

Warm blooded chameleon

Body temperature remains same when it is cold or hot outside

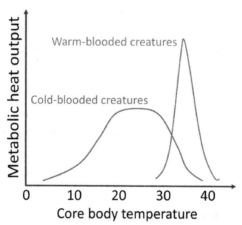

Cold blooded goose

Body temperature is dependent on whether it is cold or hot outside

Figure 11.2.   Are you warm or cold blooded? (created by S.K. Mohanakrishnan).

Figure 11.3.   Metabolic heat output regulation: warm-blooded versus cold-blooded (created by D. Ting).

Dayaks) prosper. It takes some adaptation for Homo sapiens to adopt the higher-than-comfortable humidity at a somewhat-higher-than-ideal temperature of the rainforest. As far as thermal comfort is concerned, even the native rainforest habitants prefer slightly cooler and moderately less-humid conditions. It is important to note that the lush vegetation significantly cool the surroundings as well as the mind. Therefore, residing in the middle of the forest can be notably more comfortable, but not so in urban or suburban areas where the forest is largely cleared.

Desert plants: Abundant solar energy alone, without moisture, is an unwelcome environment. Such is the case in the desert, where the sun is scorching in the scarcity of moisture. Hostile as it is, some resilient species are equipped to survive in these thermally harsh conditions. Succulent plants such as cacti and aloe can absorb large quantities of water during the odd occasions when it rains in the desert. The precious water is stored and preserved by their waxy, waterproof skin, preventing them from dying of thirst during the extended drought season between rains.

Wintery greens: What about in the cold? The most obvious plants are evergreens, most well known for making Christmas picturesque. Like an igloo, many cold-weather plants benefit from the snow for providing a layer of good insulation [Jonas et al., 2008]. Snow does not just shield the leaves from the frigid wind, but the snow on the ground also insulates the roots. In the lack of sunlight, these plants go through a sort of hibernation, ready to spring forth once the land starts to thaw.

Desert creatures: And the living creatures? For the arid desert, the extra large ears of the Fennec fox (see Fig. 11.4) keep it cool in the Sahara Desert of North Africa [Maloiy et al., 1982]. The thirst of this small carnivore is satisfied by its preys. What about in the Australian Outback, where creatures may never see a rain puddle through their entire life? Dry as it may be, there is moisture in the desert air. Thorny devils condense dew from the cool night air on the scales of their skin [Withers, 1993]. The condensate is drawn to the mouth via capillary action.

Figure 11.4. How some creatures keep cool and hydrated in an arid environment (created by S. Akhand). Fennec fox has extra large ears to keep its cool. Thorny devil uses its unique skin to condense moisture from the dry desert air to quench its thirst.

Figure 11.5. A warm-blooded creature with adjustable body temperature (created by S. Akhand).

The phenomenal camels (Fig. 11.5) definitely deserve some undivided attention; after all they faithfully serve humans in one of the most uncompromising environments. How do they do it? First and foremost, camels are no ordinary warm-blooded creatures, they can adjust their body temperature over a range that kills feeble human beings [Schmidt-Nielsen et al., 1956]. Camels have a special heat exchange mechanism that keeps their brain cool under the desert sun, though there remains some debate regarding under what conditions this special feature kicks in [Schroter et al., 1989; Mitchell et al., 2002]. Conservation and storage of the invaluable water is the key. To survive a desert excursion for up to 14 days without water, typical water storage would not suffice. Camels effectively store water, in the form of fat, in their single or double humps. As needed, the fat is broken down into water to quench their thirst. And, do not pee in the desert if you can. Since all mammals are required to urinate, camels conserve water by doing this minimally and with minimal quantity.

To survive extreme cold, warm-blooded creatures resort to snow and ice to survive. Figure 11.6 shows how a seal and a flock of penguins fight the bitter cold. The seal hides from the frigid wind in a hole surrounded by ice and snow. Zero-degree

Figure 11.6. How some warm-blooded creatures survive extreme cold (created by S. Akhand). Freezing snow and ice water are relatively warm in extremely frigid weather, and thus, the sea lion takes advantage of them. Penguins enjoy the warmth of good company.

water, in this the extreme weather condition, is like a hot tub. On the other hand, penguins take advantage of the warmth from fellow comrades.

11.4 NATURE-INSPIRED THERMAL COMFORT ENGINEERING

Swimming in frog strokes is but one of the multitude of lessons that can be learned from nature. Shaolin kung fu made its reputation based on the many potent movements of tigers, cranes, leopards, snakes, and dragons. There remains much room for us to benefit from the bountiful, beautiful, and intelligent designs in nature. With respect to nature-inspired thermal comfort engineering, passive actuators commanded by ambient weather conditions can be implemented on building envelopes. Poppinga et al. [2018] conveyed that serious energy savings can be achieved if the many actuators associated with the building envelope for controlling thermal comfort are passively and directly triggered and/or powered by changes of the environmental conditions. With the direct link between the environment and the actuators, the system becomes autonomous. One obvious present challenge is cost. Knowing how technologies have escalated in recent years, it will not be long before these autonomous actuators become affordable and widespread. Fecheyr-Lippens and Bhiwapurkar [2017] focused specifically on applying biomimicry to design building envelopes that lower energy consumption in a hot and humid climate. To mention but one yet-to-be implemented possibility, the Hercules beetle changes its color as dictated by the environment. When it is dry, its green color corresponds to pores that are filled with air. The pores are filled with water when it is humid, and its color changes to black.

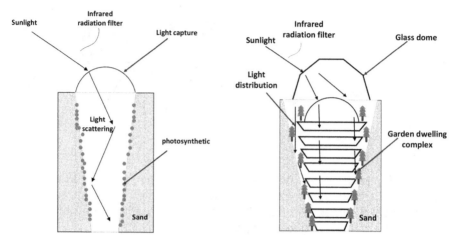

Figure 11.7. A desert building design based on fenestraria aurantiaca and frithia pulchra (created by X. Wang).

One somewhat well-established building design that is based on biomimicry is the Eastgate Centre, Harare, Zimbabwe. How cool is a termite mound such as those shown on the book cover? It is so cool that it is untouchable by forest fires. The key to termite-mound-based buildings is effective natural convection, from the ground level all the way up through the roof. For a big, tall building, the savings in mechanically moving the ventilating air are immerse. Can the same idea be adopted in a desert environment? It is absolutely possible. Figure 11.7 depicts the design of a desert building rooted deep into the ground. This is imitating desert or stone plants, fenestraria aurantiaca and frithia pulchra, where a major portion of the plant is under the ground, away from the blazing desert sun and wind.

With HVAC accounting for roughly 50% of a building's total energy consumption, savings along these lines can drastically mitigate the many challenges associated with future energy needs and environment protection. The initial costs and implementation are yet to be ironed out. Martín-Gómez et al. [2019] detailed an extensive list of possibilities regarding adopting the intelligent designs in animals to advance building energy performance. Badarnah and Kadri [2015] detailed a methodology for the generation of biomimetic design concepts. A platform of investigative tools for biomimicry as a new approach for energy-efficient building design has recently been proposed by Chayaamor-Heil and Hannachi-Belkadi [2017]. It is understandable that the passion of some of these researchers could have stretched the possibilities a little too far. Nonetheless, there is a lot out there for us to attempt.

PROBLEMS

Problem 11.1

Explain three key unique HVAC-related features of Esplanade—Theatres on the Bay, Singapore. If the building envelope were smooth like a melon, instead of that of a durian, how would this affect the HVAC energy usage?

Problem 11.2

Deduce the pressure difference due to stack effect, ΔP_s, of the Eastgate Centre, Harare, Zimbabwe. Estimate the volume flow rate of air up a 2-m-diameter air shaft.

Problem 11.3

Explain how the fat-water conversion works in camels.

Problem 11.4

Shark skin has been found to be intelligently designed to minimize friction (drag). How much pumping power can be saved if the inner surface of the water distribution pipe of a hotel copies the texture of the shark skin? Assume that the piping networks consists of 500 m of 1.27-cm iron pipe with an average water velocity of 0.5 m/s.

REFERENCES

L. Badarnah, U. Kadri, "A methodology for the generation of biomimetic design concept," Architectural Science Review, 58(2): 120–133, 2015.

N. Chayaamor-Heil, N. Hannachi-Belkadi, "Towards a platform of investigative tools for biomimicry as a new approach for energy-efficient building design," Buildings, 7(19): 1–18, 2017.

R.P. Clark, "Human skin temperature and its relevance in physiology and clinical assessment," in E.F.J. Ring, B. Phillips (Editors), Recent Advances in Medical Thermology, Springer, Boston, 1984.

D. Fecheyr-Lippens, P. Bhiwapurkar, "Applying biomimicry to design building envelopes that lower energy consumption in a hot-humid climate," Architectural Science Review, 60(5): 360–370, 2017.

E.N. Hey, G. Katz, "The optimum thermal environment for naked babies," Archives of Disease in Childhood, 45: 328–334, 1970.

T. Jonas, C. Rixen, M. Sturm, V. Stoeckli, "How apline plant growth is linked to snow cover and climate variability," Journal of Geophysical Research, 113: 1–10, 2008. doi:10.1029/2007JG000680

C.L. Lim, C. Byrne, J.K.W. Lee, "Human thermoregulation and measurement of body temperature in exercise and clinical settings," Annals Academy of Medicine, 37(4): 347–353, 2008.

Z. Lin, S. Deng, "A study on the thermal comfort in sleeping environments in the subtropics – Developing a thermal comfort model for sleeping environments," Building and Environment, 43: 70–81, 2008.

G.M.O. Maloiy, J.M.Z. Kamau, A. Shkolnik, M. Meir, R. Arieli, "Thermoregulation and metabolism in small desert carnivore: the Fennec fox (Fennecus zerda) (Mammalia)," Journal of Zoology, 198: 279–291, 1982.

C. Martín-Gómez, A. Zuazua-Ros, J. Bermejo-Busto, E. Baquero, R. Miranda, C. Sanz, "Potential strategies offered by animals to implement in buildings' energy performance: theory and practice," Frontier of Architectural Research, 8: 17–31, 2019.

D. Mitchell, S.K. Maloney, C. Jessen, H.P. Laburn, P.R. Kamerman, G. Mitchell, A. Fuller, "Adaptive heterothermy and selective brain cooling in arid-zone mammals," Comparative Biochemistry and Physiology, Part B, 131: 571–585, 2002.

O. Nordahl, P. Tibblin, P. Koch-Schmidt, H. Berggren, P. Larsson, A. Forsman, "Sun-basking fish benefit from body temperatures that are higher than ambient water," Proceedings of the Royal Society B, 285: 20180639, 2018.

S. Poppinga, C. Zollfrank, O. Prucker, J. Rühe, A. Menges, T. Cheng, T. Speck, "Toward a new generation of smart biomimetric actuators for architecture," Advanced Materials, 30: 1703653, 2018.

S.A. Rommel, D.A. Pabst, W.A. McLellan, "Reproductive thermoregulation in marine mammals," American Scientist, 86(5): 440–448, 1998.

K. Schmidt-Nielsen, B. Schmidt-Nielsen, S.A. Jarnum, T.R. Houpt, "Body temperature of the camel and its relation to water economy," American Journal of Physiology, 188(1): 103–112, 1956.

R.C. Schroter, D. Robertshaw, R.Z. Filali, "Brain cooling and respiratory heat exchange in camels during rest and exercise," Respiration Physiology, 78(1): 95–105, 1989.

P. Withers, "Cutaneous water acquisition by the thorny devel (Moloch horridus: Agamidae)," Journal of Herpetology, 27(3): 265–270, 1993.

Index

CPSIA information can be obtained
at www.ICGtesting.com
Printed in the USA
BVHW040746260320
575395BV00010B/14

9 789811 201745